T0292602

Springer Theses

Recognizing Outstanding Ph.D. Research

Aims and Scope

The series "Springer Theses" brings together a selection of the very best Ph.D. theses from around the world and across the physical sciences. Nominated and endorsed by two recognized specialists, each published volume has been selected for its scientific excellence and the high impact of its contents for the pertinent field of research. For greater accessibility to non-specialists, the published versions include an extended introduction, as well as a foreword by the student's supervisor explaining the special relevance of the work for the field. As a whole, the series will provide a valuable resource both for newcomers to the research fields described, and for other scientists seeking detailed background information on special questions. Finally, it provides an accredited documentation of the valuable contributions made by today's younger generation of scientists.

Theses are accepted into the series by invited nomination only and must fulfill all of the following criteria

- They must be written in good English.
- The topic should fall within the confines of Chemistry, Physics, Earth Sciences, Engineering and related interdisciplinary fields such as Materials, Nanoscience, Chemical Engineering, Complex Systems and Biophysics.
- The work reported in the thesis must represent a significant scientific advance.
- If the thesis includes previously published material, permission to reproduce this must be gained from the respective copyright holder.
- They must have been examined and passed during the 12 months prior to nomination.
- Each thesis should include a foreword by the supervisor outlining the significance of its content.
- The theses should have a clearly defined structure including an introduction accessible to scientists not expert in that particular field.

More information about this series at http://www.springer.com/series/8790

Vlado Menkovski

Computational Inference and Control of Quality in Multimedia Services

Doctoral Thesis accepted by
Eindhoven University of Technology, The Netherlands

 Springer

Author
Dr. Vlado Menkovski
Philips Research
Eindhoven, Noord-Brabant
The Netherlands

Supervisor
Prof. Antonio Liotta
Eindhoven University of Technology
Eindhoven
The Netherlands

ISSN 2190-5053   ISSN 2190-5061   (electronic)
Springer Theses
ISBN 978-3-319-24790-8   ISBN 978-3-319-24792-2   (eBook)
DOI 10.1007/978-3-319-24792-2

Library of Congress Control Number: 2015950057

Springer Cham Heidelberg New York Dordrecht London
© Springer International Publishing Switzerland 2015
This work is subject to copyright. All rights are reserved by the Publisher, whether the whole or part of the material is concerned, specifically the rights of translation, reprinting, reuse of illustrations, recitation, broadcasting, reproduction on microfilms or in any other physical way, and transmission or information storage and retrieval, electronic adaptation, computer software, or by similar or dissimilar methodology now known or hereafter developed.
The use of general descriptive names, registered names, trademarks, service marks, etc. in this publication does not imply, even in the absence of a specific statement, that such names are exempt from the relevant protective laws and regulations and therefore free for general use.
The publisher, the authors and the editors are safe to assume that the advice and information in this book are believed to be true and accurate at the date of publication. Neither the publisher nor the authors or the editors give a warranty, express or implied, with respect to the material contained herein or for any errors or omissions that may have been made.

Printed on acid-free paper

Springer International Publishing AG Switzerland is part of Springer Science+Business Media
(www.springer.com)

To my parents, Angel who dedicated his life to his sons and Snezhana whose strength inspires me deeply.

Supervisor's Foreword

Streaming video over packet-based communication networks is as important as it is difficult. At their conception in the 1960s, the Internet protocols that we still use today were not designed for this use but were meant for asynchronous communications (such as emails) rather than time-constrained media (such as audio and video). On the other hand, videos were originally analog signals, only meant for local replay in movie theaters. Yet today the predominant Internet traffic is made of video streams in all sorts of forms, video-on-demand, IPTV, conferencing, peer-to-peer, and so forth. This requires videos to be digitized, compressed, split in packets, transmitted over lossy networks, and then reconstructed on a player. At each stage, important information is lost, which makes it difficult to push video over networks that can lose large chunks of data by way interference, congestion, latency, and whatnot.

How much bandwidth is necessary to ensure a transmission of 'good' quality? How far can we shrink a stream? How can we broadcast to a large audience in real time? It turns out that these broad questions are largely unanswered and scientists are still busy sorting out the pieces of this complex puzzle. How can we actually measure video quality, which involves forays in the mysterious realm of human perception? How do the individual stages of video transmission affect 'perceived' quality? In which ways should we re-design the network to be more video-friendly?

Despite this large knowledge gap, video services are proliferating and thanks to an engineering trick that goes under the name of 'overprovisioning.' Until we know how to best master the transmission of time-critical media over packet-switching networks, the only option we have is equip the network with an excess of capacity and servers. This practice is utterly expensive and energy-demanding and unsustainable in the longer term, considering the relentless request for video services.

As new technologies stack on top of each other, the interactions among the various pieces become evidently intricate and obscure. Some people think that attempting to master the complexity of Internet systems is a lost cause, thus advocating the 'overprovisioning' route as the only viable avenue. My line of thought goes the other way and I believe that, as scientists, we need to find ways to reduce the energy footprint of the Internet. In turn, we need to understand the mutual influence between networks and services, and study new ways to control them optimally.

The work that Vlado Menkovski describes in this book makes a number of breakthroughs at the intersection between networks, video services, human perception, and computational intelligence. He shows various ways in which machine learning may be used to assess the quality of current media streaming methods and to make them more efficient. While reading the book, at some point you will realize why 'quality of experience management' is not merely a control problem—it is a matter of prediction. I truly hope that you will enjoy reading this book as much as I have enjoyed working with Vlado, and that you will find it not only useful but also full of inspirational ideas.

Eindhoven, The Netherlands   Prof. Antonio Liotta
October 2015   Chair of Communication Network Protocols

Preface

Quality is the degree of excellence we expect of a service or a product. It is also one of the key factors that determine its value. For multimedia services, understanding the experienced quality means understanding how the delivered fidelity, precision, and reliability correspond to the users' expectations. Yet the quality of multimedia services is inextricably linked to the underlying technology. It is developments in video recording, compression, and transport, as well as display technologies that enable high quality multimedia services to become ubiquitous. The constant evolution of these technologies delivers a steady increase in performance, but also a growing level of complexity. As new technologies stack on top of each other the interactions between them and their components become more intricate and obscure. In this environment optimizing, the delivered quality of multimedia services becomes increasingly challenging. The factors that affect the experienced quality, or Quality of Experience (QoE), tend to have complex nonlinear relationships. The subjectively perceived QoE is hard to measure directly and continuously evolves with the user's expectations. Faced with the difficulty of designing an expert system for QoE management that relies on painstaking measurements and intricate heuristics, we turn to an approach based on learning or inference. The set of solutions presented in this work rely on computational intelligence techniques that do inference over the large set of signals coming from the system to deliver QoE models based on user feedback. We furthermore present solutions for inference of optimized control in systems with no guarantees for resource availability. This approach offers the opportunity to be more accurate in assessing the perceived quality, to incorporate more factors, and to adapt as technology and user expectations evolve. In a similar fashion, the inferred control strategies can uncover more intricate patterns coming from the sensors and therefore implement farther-reaching decisions. Similarly to biological systems, this continuous adaptation and learning make these systems more robust to perturbations in the environment, longer lasting accuracy, and higher efficiency in dealing with increased complexity. Overcoming this increasing complexity and diversity is crucial for addressing the challenges of future multimedia system. Through experiments and simulations, this work demonstrates that adopting an approach of learning can improve the subjective and objective QoE estimation, enable the implementation of efficient and scalable QoE management as well as efficient control mechanisms.

Contents

Contents

Chapter 1
Introduction

Abstract The visual senses dominate the sensory input, accounting for 80 % of perceptual information. This makes video an attractive medium for high density information services. As video enabled services become more present in our lives, our expectations about their performance and reliability is being set. In order to meet customer's expectations, service providers need to be able to deliver increasingly more demanding services with higher quality standards. This development delivers a high toll on maintenance costs and requires frequent upgrades of available resources. Moreover, the upgrade of some wired and wireless transmission technologies is becoming more challenging as technologies are reaching some physical limits. In this situation the need for smarter management strategies is evident as traditional management approaches such as over-provisioning offer little to improve the utilization of the resources. Efficient management of networked services requires understanding of the relationship between different available resources, i.e. computational, storage, network throughput and the delivered quality. However, video-enabled services are operating on a vast diversity of terminal devices, encoding and transmission systems. Motivated by these challenges, this thesis proposes an approach for efficient management of multimedia services. It presents a QoE aware framework for network management that incorporates computational intelligence methods to deal with the evolving complexities in the multimedia systems, and introduces a novel psychometric method that deals with the difficulties of subjective measurements.

1.1 QoE Management Framework

This work proposes an approach for end-to-end management of quality of multimedia services as a framework on top of multitude of underlying technologies [1–4]. At one end is the service provider with the content ready to be served and on the other is the user with its unique characteristics and expectations regarding the content [5, 6]. The framework is designed as a control loop over a general-purpose multimedia system with the goal of matching the properties of the content to the expectations of the consumer, while accounting for the available resources and characteristics of the encoding and the transport systems.

© Springer International Publishing Switzerland 2015
V. Menkovski, *Computational Inference and Control of Quality in Multimedia Services*, Springer Theses, DOI 10.1007/978-3-319-24792-2_1

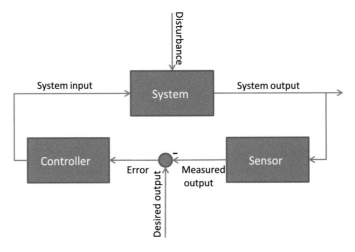

Fig. 1.1 A feedback loop in a control system

As illustrated in Fig. 1.1, a negative feedback control loop consists of three units: the controller, the sensor and the system under control. The sensor measures the system output and compares it with the desired one. The difference is fed into the controller that inputs a control strategy to the system in order to minimize the measured difference. Similarly, the multimedia system controller issues different management strategies based on the measured performance of the system (Fig. 1.2). The

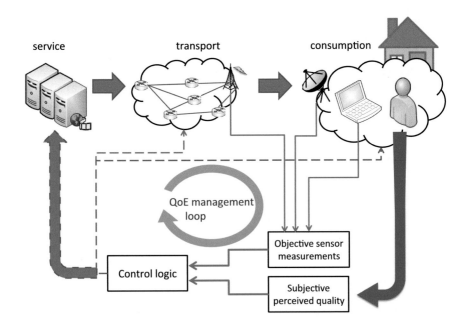

Fig. 1.2 Control loop in a multimedia system

measured value is the subjective QoE perceived by the user and the objectively measured performance by the system components [7]. This value is compared with the desired level of performance and the difference is sent to the controller. The controller can then manage the different system components, allocate necessary resources, execute admission control, or implement other control strategies to achieve the desired level of performance.

This approach offers a viable way to incorporate the large number of factors that affect the quality into the decision process of the controller. It provides for a way to continuously learn and improve the management process based on measurements of the performance and subjective user feedback [8]. In this manner the system maintains a high level of performance with minimum cost in the changing environment. This results in a better utilization of the available resources and a user-centric based management [9–14].

1.2 QoE Definition

In the framework presented here, the system output (or the main performance metric) is the subjectively perceived quality. But how do we understand the term 'quality' in regards to multimedia services?

According to the dictionary, the meaning of quality is: 'The degree of excellence' [15] (of a product, service or activity). Quality as a phenomenon has been examined in many disciplines such as philosophy, business and engineering. More specifically quality involves perception, but also expectations. Some definitions refer to it as a subjective phenomenon: "The feeling of high quality occurs when perception exceeds expectation; the feeling of low quality occurs when perception does not meet expectation [16]." Other definitions focus on more objective, measurable factors: "Degree to which a set of inherent characteristics fulfils requirements." where requirement is defined as need or expectation [17]. In any case, quality is connected with either objective or subjective expectations. More precisely quality is evaluated in regards to the objectively measured or subjective perceived performance.

For services such as telephony, computer networks, and including voice services built on top of them a commonly used metric for quality is the Quality of Service (QoS). QoS defines a set of requirements that need to be met in order for the service to be considered of high quality. These requirements are objectively measurable values and consider performance factors, such as latency and errors in the network. The possibility to set these well defined QoS requirements is enabled by the good understanding of the compression and transmission factors and the subjective perception of speech. However, with the introduction of new types services and varied content, all delivered on a plethora of different devices, understanding 'subjective' perception becomes challenging.

The need to better communicate the service quality has created a need for a precise quality metric. However, instead of continuously expanding and adapting the QoS requirements, the choice was made to introduced a new metric: the Quality of Experience (QoE) [18]. This metric is better suited for the task, because it is a subjective metric, which captures the effect of all the factors that contribute to the subjective experience.

QoE appears in the literature by different definitions, but generally it is agreed that "QoE measures the quality experienced while using a service" [19]. However, other definitions, such as the one from the ITU-T Focus Group on IPTV (FG IPTV) [20] avoid using the term quality. The FG IPTV defines QoE as the overall acceptability of an application or service, as perceived subjectively by the end-user. The definition of the European Network on QoE in multimedia systems and services is *'QoE is the degree of delight or annoyance of the user of an application or service. It results from the fulfilment of his or her expectations with respect to the utility and/or enjoyment of the application or service in the light of the user's personality and current state'* [18]. The relation to the subjective perception of the user and its expectations is clearly evident as the defining characteristic of the metric.

Defining QoE is the initial step. In order to successfully build an efficient management system we need to first understand which factors affect it and how [21, 22]. Next we need to understand the relationship between the available resources in the system and those factors [3, 4, 11]. Finally we need to know how to develop control strategies that utilize the available resources in such a manner as to maximize the delivered QoE [23, 24].

1.3 Factors that Affect QoE

QoE is a metric that captures the degree to which our expectations about the service have been met. But, how do we form expectations of a video service and which are the factors that affect it?

We perceive multimedia stimuli first with our senses, and ultimately through the cognitive processes [25] in the brain. Naturally, the characteristics of the human visual and auditory systems are intrinsically linked to the expectations from multimedia services. The auditory system can detect sound in a specific range of frequencies. The typical hearing range for a human is between 20 Hz and 20 KHz [26]. Therefore, reproduction of sound outside of this range will add no additional quality or value to the service; it will only spend more resources. The visual system has elevated sensitivity to contrast in the range of 0.5–16 cycles per degree of visual angle and drops of abruptly on higher frequency [27]. This means that a 1080p HDTV reproduction viewed at a distance between 3 and 4 screen heights can generate patterns on our retina with up to 28 and 37 cycles per degree respectively. This is significantly above the threshold limit for most viewers [28], so increasing the resolution more at this distance will be of little utility.

These examples demonstrate that even though we intuitively consider fidelity of the video and audio as the most significant factor in quality, this has to be taken into the context of the characteristics of the Human Visual System (HVS). Improving certain aspects of fidelity could be without any benefit. On the other hand, limited loss of fidelity can be just as imperceptible as a flawless reproduction, while delivering significant benefits in terms of resource utilization.

The Human visual and auditory system is not only characterized by a hearing range and a contrast sensitivity range. It is actually very complex and not fully understood. As more characteristics are being discovered the more we can use this knowledge to optimize multimedia services. Some of these limitations are commonly used to improve coding efficiency. For example high spatial frequencies are perceived achromatic [29]. So the amount of data that conveys the colour of the image can be safely reduced in respect to the amount of data that carries the luminance of the image. Other characteristics, such as the masking effects can be used to cover noise in images. If the noise is superimposed over a region of patterns with high contrast, it is significantly less perceptible than over a uniformly coloured region.

Video compression techniques benefit from the varying sensitivity to different ranges of spatial frequency. The video is first transformed into the frequency domain using a discrete cosine transform (DCT) [30]. Then different parts of the image can be encoded with different precision. The last step is known as quantization, whereby high-frequency coefficients are more coarsely quantized than low-frequency coefficients. This is referred to as a lossy compression method. Even though lossy compression can deliver significant benefits in reducing the size of the video, it also causes loss in quality. Quantization causes artefacts such as blockiness, particularly in heavily compressed video, which degrade the QoE.

Further artefacts appear in a video due to modern compression techniques, such as blur, colour bleeding, ringing, false edges and jagged motion [28]. Some of them are present due to spatial compression techniques, which compress individual images. Others are present due to temporal compression method, which reduces the redundancy over multiple images.

Another factor that introduces artefacts is transmission errors [31, 32]. When a packet of video data is lost, has errors or does not arrive on time the video decoder can freeze the video playback or compensate by using some concealment method. Usually this means interpolating neighbouring pixels in space and in time [33]. However, this often results in very noticeable artefacts [34, 35].

When transmission protocols are used to guarantee delivery via retransmission, the lack of network resources leads to delays and freezes in playback. This is a very important factor in the overall experience of the service and can have the most significant impact [36]. It is also established that impairments such as freezes and errors have higher impact on the QoE as their amplitude and frequency increase [37].

Adaptive video streaming technologies allow for reducing the fidelity of the signal in order to avoid freezes. This technique attempts to improve the delivered QoE in cases of restricted resources by downloading the video at lower bit-rates. In this case predicting the right bit-rate level is important as the changes of bit-rate levels during playback have shown to be an impairment on its own. The size of the impairment is proportional to the frequency and amplitude of the change [38].

Naturally, the audio quality is a key factor in the QoE of multimedia services. In fact, audio quality is even more important factor than video quality [28]. Audio compression also benefits from the characteristics of the auditory system. Lossy audio compression methods cause audio artefacts in a similar fashion to the video compression methods. However, audio compression and encoding requires significantly less resources than video and, because of its importance, its resources are rarely restricted. This often shifts the management focus on the video aspect.

Nevertheless, other more general aspects are also important factors. One such example is the audio and video synchronization. The investigation of media synchronization in [39] concludes that the effect of unsynchronized audio on the QoE depends on the type of content. For some types of content as head and shoulders news broadcast, it has a massive effect. However, for other content the viewers can demonstrate more tolerance. In another investigation of audio and video correlation and lip synchronization Mued et al. conclude that the effects on the perceived quality from audio-video misalignment are different when the content is of a passive or an active communication [40].

Depending on the type of service, there could be other types of impairments such as start-up delays or loss in responsiveness. Overall the factors that affect the QoE are noticeable and are not expected by the viewer. We use our eyes and ears to collect the information from the outside world, but it is the brain that forms our perceptions [41]. The cognition process in the human brain is not understood well, however we know that we do not need all the details to recognize a pattern. Our sensors are designed in this manner, working in restricted ranges. The rest of the details are conceptualized by the cognition process. Nevertheless, with fewer details, the brain needs to work harder to compensate. Sometimes we are willing to do that, because we are watching an old family movie on an outdated technology. But, other times, when we are watching a video on our new and costly mobile device our expectations are high, so the delivered QoE needs to reflect that. As this might be an insurmountable task, measuring QoE in relative terms can be a better solution than attempting to make an inaccurate absolute metric. Pursuing this approach we have developed a method for measuring subjective video quality, which estimates the utility of the system resources in terms of the delivered QoE [42, 43].

1.4 Resources Versus Quality

We have seen that a significant amount of factors contribute to the QoE. However, there is also a clear relationship between these factors and the available resources. Most types of artefacts can be efficiently masked if we can encode the video with enough bits, which also need to be transported by the network accurately and on time. Unfortunately, in any real engineered system the available resources are limited. We can only serve a limited number of users, the network can only transport a limited number of bits per second, and finally storage space and computational power are equally limited. So in order to efficiently manage the system we need to understand the relationship between the allocated resources and the resulting QoE.

The audio and video fidelity have a clear relationship with the available resources. The more bits are used in the encoding process the more accurate the decoded audio and video will be. Uncompressed video contains no encoding degradation, but it requires very large storage space and is not suitable for transmission over a restricted network channel [44]. Compression can be 'lossless' where the signal can be exactly reproduced or 'lossy' that introduces loss in the fidelity of the signal. Since high definition multimedia content requires significant compression rates (50:1 [45]), lossy encoding techniques present an attractive choice. This is further supported by the leniency of the human visual and auditory system to certain type of distortions. So, the optimal quality is usually achieved in a balanced combination of the parameter values that results in minimal use of resources and a satisfactory level of accuracy.

Each digital video encoding process produces video streams with specific bit-rates. The bit-rate is directly linked to the quality of the video stream and most encoders accept a bit-rate setting as input. It can be either set as a soft (indication) or as a hard (constraint) limit on the encoder in constant bit-rate encoding. In variable bit-rate encoding (VBR), the indicator is usually a quality setting. Therefore, the stream is with constant quality rather than having a constant bit-rate. Based on this setting and the complexity of the video, the encoder compresses the video with a certain average bit-rate. Therefore the bit-rate required to encode the video depends on the type of encoding algorithm, the complexity of the video and the desired quality. Since transport throughput is also a limited resource, the video bit-rate needs to be adjusted accordingly in order to meet the transport network constraints. This is commonly achieved by compressing the video with constant bit-rate encoding (CBR). Typically, MPEG-like algorithms will introduce increasingly larger amounts of artifacts (such as blockiness and blurriness) as the bit-rate is reduced [28]. In other words the video data will be more coarsely quantized in the frequency domain, which will lead to blockiness in the decoded video. The encoder attempts to limit the blockiness effect on low spatial frequencies of the video, which are less perceptible to the viewers. However, very constrained compressions result in highly visible artefacts. Another type of artifact due to encoding is blurriness. This one arises from inadequate temporal fidelity of the encoded video [46].

The loss of fidelity can originate from the pre-encoding process as well. The spatial resolution of the digital video is one of the key factors for the size of the video after encoding. The recording equipment usually has much higher resolution that what can be practically used in video streaming applications. Particularly for mobile devices the resolution needs to be adjusted to the limitations of the devices in screen resolution, computational power or network throughput. It is common to downscale the resolution of the original video before encoding, because it reduces the encoding computation time as well as the size of the resulting video. In cases where the target screen resolution is lower than the input video this pre-encoding process is only beneficial. On the other hand, restricting only the bit-rate can degrade the video more, add more computational on the encoder and on the decoder for downscaling.

In addition to the spatial resolution decrease, there is temporal resolution decrease, or decrease in frame-rate. Frame-rate is usually kept to less than 30 frames per second due to the characteristics of the human visual systems. However frame-rate acceptability depends on the type of content [6, 47, 48]. Certain types of content that have low mobility and small spatial resolution, frame-rates as low as 10 frames per second can be acceptable. This is particularly useful for very low bit-rate channels in mobile environments where lower frame-rates help achieve the required low throughput. Similar pre-encoding can be implemented in the audio, when sampling rate and sampling frequency are downscaled.

Making the right decisions during the encoding process is key to minimizing the amount of delivered artifacts. Rate distortion theory provides a the theoretical foundation for this problem. The theory is a branch of information theory and deals with lossy data-compression [49]. Based on this theory many rate distortion optimization (RDO) methods have been developed and are commonly incorporated in the decision process of multimedia encoders [50].

Another reason for occurrence of artefacts is errors incurred during transmission. The amount of degradation in quality due to such errors is not easy to estimate due to the very nature of video compression [51]. The removal of temporal redundancy in the video leads to propagation of the errors in multiple frames. This can be constrained by adding more I-frames or reference frames that do not require frame from other data to be compressed. However, increasing the frequency of the I-frames decreases the efficiency of the compression. Another approach to protect the data stream is to use forward error correction techniques [52]. These techniques add redundant data to the stream, which allows recovery of a limited number of bits in case of errors or loss during transmission. Selecting the appropriate amount of redundant data can be achieved by applying RDO to this problem as well [53].

When transport protocols ensure delivery via packet retransmission the same mechanisms can cause delays and hence lower throughput. The network throughput can also fluctuate, causing jitter in the arrival time of the packets. This can result in difficulties for streaming of video as the decoder cannot wait for late video packets. Buffering is used to compensate for the effects of delay and jitter. However, in order to compensate for high variation in packet arrival time significantly large buffer is necessary. This imposes high memory requirements on the client but it also has a direct effect on QoE, because it increases the start-up time of the playback. This effect can be sometimes more damaging to QoE than lowering the bit-rate.

In this thesis we demonstrate a number of objective and subjective QoE measurement techniques applied on a limited range of factors. We have developed implementations on existing techniques and certain extensions where the state-of-the-art did not produce desired results. The results from these measurements deliver valuable information on the effects of these factors on QoE.

Nevertheless, in the effort to capture a fuller picture of the QoE of an operating multimedia system, it becomes evident that there are a large number of factors in play and there are an equally large number of parameters in the system that contribute to this factors. The relationship between the parameters, resources and technology in the system can be complex and highly non-linear [54]. So, there is a clear need to deal with this complexity and uncover the relationship between the system intricacies and the QoE [5].

1.5 Handling the Complexity of QoE Modelling

Our ambition is to create a framework for QoE-aware management of multimedia services. In order to do that we need to be able to understand how our management decisions are affecting the QoE. Yet, as QoE is multifaceted and has complex relationships with the available resources in the system, its modelling presents a challenge.

Many subjective studies of certain aspects of QoE (including our own) have been executed. Most of them are focused on the efficiency of the encoding [55], while others on the effects of specific errors on the quality. However, considering all the factors that affect quality in a subjective study would be an insurmountable task.

A more tractable way to deal with the multitude of factors that affect QoE is to use computational techniques. Computational intelligence and, more concretely, Machine Learning techniques offer the ability to correlate a vast amount of parameters with each output metric. They can discover complex interdependencies and detect non-obvious patterns. QoE models developed by combining objectively and subjectively measurable factors can deliver much better understanding of the delivered QoE to the viewer than by just looking at individual parameters such as bit-rate or video resolution.

In this thesis we present a system that collects a multitude of measurements from a multimedia system and correlates this information with subjective feedback from the users. The models delivered from the correlation can be further used to estimate the performance of the system over a longer period of time.

On-line learning techniques can be used to continuously adapt these models and deliver an accurate estimation of QoE, even in a continuously changing environment [56–58]. The understanding of QoE can deliver an efficient longer term management cycle of monitoring, evaluation and provisioning. Despite that, when faced with active control or short-term management decisions we need to understand the effect of each decision on the QoE. For this type of management instead of inference and modelling we need to move on to optimal control strategies.

In the following chapters we present description of a QoE management framework that addresses the challenge of complexity and adapts in an on-line fashion. We also present an approach of QoE-aware active control of multimedia systems, where short term decisions are made in correspondence with the fluctuations in the available resource.

1.6 Learning Versus Deterministic Design

Management of networked services typically means provisioning enough resources and allocating them appropriately. However, as certain resources are shared over the systems their availability varies over time. For many applications, making real-time decisions on the available resources makes the difference between delivering high quality service and failing to do so.

The usual approach in developing an efficient controller for real-time management is to design a suitable heuristic (or a rule based system) that will take appropriate decisions based on the state of the system. This approach requires a thorough understanding of the effects of the decisions on the performance of the system. As complexity in the system grows the design of efficient heuristics becomes more challenging and more expensive. As the system evolves rules become outdated [59].

In some areas, such as video encoding and video streaming RDO methods have been implemented to optimize the trade-off between quality and resources. Even though these methods have sufficient theoretical basis, practically the models that they rely on to calculate the rate and distortion do not fully capture the complexity of different video sources [45]. Furthermore, with the growth in the complexity of the systems, the interdependencies between the decision are not fully taken into account [60].

In contrast to this methodology, in this thesis we present an approach of 'learning' optimal strategies rather than 'designing' them. A computational intelligence technique based on reinforcement learning is used to discover the longer term utility of the decision, given the state of the system, and develop an optimal strategy.

This technique relies on previous techniques for modelling QoE and on well-established methods for reinforcement learning, offering an approach for designing system control that is scalable and adaptable to the changes in the environment.

1.7 Main Contributions

The focus of this work is developing methods for efficient management and control of delivered quality in multimedia services.

The main challenges facing this goal are understanding the different factors that affect the quality, the growing complexity in the interaction of those factors and the effect of the management decision on the quality.

In order to understand the delivered quality, we have defined QoE as its metric, discussed the factors that affected it and the resources that relate to these factors.

In Chap. 2 we continue to present a discussion on objective QoE methods and our implementations and developments of supporting techniques. Objective QoE methods are a cost effective way to measure the factors the contribute to the delivered quality. Their use is widespread, and they correlate in varying degree with the subjective QoE.

To understand the delivered QoE thoroughly, we turn to the subjective QoE methods in Chap. 3. This chapter presents a discussion on the existing subjective QoE methods and their drawbacks. Furthermore, it introduce a novel video subjective method based on psychometric evaluation that addresses many of these challenges.

Multimedia delivery systems are typically complex and their successful management requires a broader approach. In Chap. 4 we present our framework for QoE monitoring and provisioning that learns how to model all available measurements into a QoE value. Moreover, we provide a solution for calculating the remedies in systems where the measured values are not satisfactory.

In following chapter (Chap. 5) we present our approach for real-time management or control of a multimedia system that infers the control logic based on the measured QoE.

The work on objective and subjective QoE models builds a basis for the QoE management and the QoE active control framework. These two frameworks offer a method for efficient management and control of the quality in multimedia services faced with growing complexity and continuous evolution of both the user expectations and the underlying technologies.

References

1. F. Agboma, A. Liotta, Quality of experience management in mobile content delivery systems. J. Telecommun. Syst. (special issue on the Quality of Experience issues in Multimedia Provision) **49**(1), 85–98 (Springer, 2012). doi:10.1007/s11235-010-9355-6
2. V. Menkovski, G. Exarchakos, A. Liotta, A. Cuadra Sánchez, Quality of experience models for multimedia streaming. Int. J. Mob. Comput. Multimed. Commun. **2**(4), 1–20 (2010). doi:10.4018/jmcmc.2010100101, ISSN:1937-9412, www.igi-global.com/ijmcmc/
3. D. Constantin Mocanu, G. Santandrea, W. Cerroni, F. Callegati, A. Liotta, Network performance assessment with quality of experience benchmarks, in *Proceedings of the 10th International Conference on Network and Service Management*, Rio de Janeiro, Brazil, 17–21 Nov 2014 (IEEE)
4. M. Torres Vega, S. Zou, D. Constantin Mocanu, E. Tangdiongga, A.M.J. Koonen, A. Liotta, End-to-end performance evaluation in high-speed wireless networks, in *Proceedings of the 10th International Conference on Network and Service Management*, Rio de Janeiro, Brazil, 17–21 Nov 2014 (IEEE)
5. F. Agboma, A. Liotta, Addressing user expectations in mobile content delivery. J. Mob. Inf. Syst. (Special issue on Improving Quality of Service in Mobile Information Systems, Services and Networks) **3**(3), 153–164 (IOS Press, 2007)
6. A. Liotta, L. Druda, G. Exarchakos, V. Menkovski, Quality of experience management for video streams: the case of Skype, in *Proceedings of the 10th International Conference on Advances in Mobile Computing and Multimedia*, Bali, Indonesia, 3–5 Dec 2012 (ACM). http://dx.doi.org/10.1145/2428955.2428977
7. V. Menkovski, G. Exarchakos, A. Cuadra-Sanchez, A. Liotta, Measuring quality of experience on a commercial mobile TV platform, in *Proceedings of the 2nd International Conference on Advances in Multimedia*, Athens, Greece, 13–19 June 2010 (IEEE)
8. V. Menkovski, G. Exarchakos, A. Liotta, Machine learning approach for quality of experience aware networks, in *Proceedings of Computational Intelligence in Networks and Systems*, Thessaloniki, Greece, 24–26 Nov 2010 (IEEE)
9. K. Yaici, A. Liotta, H. Zisimopoulos, T. Sammut, User-centric quality of service management in UMTS, in *Proceedings of the 4th Latin American Network Operations and Management Symposium (LANOMS'05)*, Porto Alegre, Brazil, 29–31 Aug 2005 (Springer)
10. F. Agboma, A. Liotta, User-centric assessment of mobile content delivery, in *Proceedings of the 4th International Conference on Advances in Mobile Computing and Multimedia*, Yogyakarta, Indonesia, 4–6 Dec 2006
11. F. Agboma, A. Liotta, Managing the user's quality of experience, in *Proceedings of the Second IEEE/IFIP International Workshop on Business-Driven IT Management (BDIM 2007)*, Munich, Germany, 21 May 2007 (IEEE)

12. F. Agboma, A. Liotta, QoE-aware QoS management, in *Proceedings of the 6th International Conference on Advances in Mobile Computing and Multimedia*, Linz, Austria, 24–26 Nov 2008
13. V. Menkovski, A. Oredope, A. Liotta, A. Cuadra-Sanchez, Predicting quality of experience in multimedia streaming, in *Proceedings of the ACM*, 7 Dec 2009. ISBN:978-1-60558-659-5, http://dl.acm.org/citation.cfm?id=1821766
14. V. Menkovski, G. Exarchakos, A. Cuadra-Sanchez, A. Liotta, Estimations and remedies for quality of experience in multimedia streaming, in *Proceedings of the 3rd International Conference on Advances in Human-oriented and Personalized Mechanisms, Technologies, and Services*, Nice, France, 22–27 Aug 2010 (IEEE)
15. Quality—definition and more from the free merriam-webster dictionary. http://www.merriam-webster.com/dictionary/quality
16. N. Kano, N. Seraku, F. Takahashi, S. Tsuji, Attractive quality and must-be quality. J. Jpn. Soc. Qual. Control **14**(2), 39–48 (1984)
17. D. Hoyle, *ISO 9000 Quality Systems Handbook-updated for the ISO 9001:2008 standard* (Routledge, 2012)
18. P. Le Callet, S. Möller, A. Perkis (eds.), Qualinet white paper on definitions of quality of experience, in *European Network on Quality of Experience in Multimedia Systems and Services (COST Action IC 1003)*, Lausanne, Switzerland, Technical Report 1.1, June 2012
19. R. Jain, Quality of experience. IEEE Multimed. **11**(1), 95–96 (2004)
20. A. Takahashi, D. Hands, V. Barriac, Standardization activities in the ITU for a QoE assessment of IPTV. IEEE Commun. Mag. **46**(2), 78–84 (2008)
21. V. Menkovski, A. Liotta, Adaptive psychometric scaling for video quality assessment. J. Signal Process. Image Commun. **26**(8), 788–799 (Elsevier, 2012). http://dx.doi.org/10.1016/j.image.2012.01.004
22. F. Agboma, M. Smy, A. Liotta, QoE analysis of a peer-to-peer television system, in *Proceedings of the International Conference on Telecommunications, Networks and Systems*, Amsterdam, The Netherlands, 22–24 July 2008
23. M. Torres Vega, D. Constantin Mocanu, R. Barresi, G. Fortino, A. Liotta, Cognitive streaming on android devices, in *Proceedings of the 1st IEEE/IFIP IM 2015 International Workshop on Cognitive Network and Service Management*, Ottawa, Canada, 11–15 May 2015 (IEEE). http://www.cogman.org
24. V. Menkovski, A. Liotta, Intelligent control for adaptive video streaming, in *Proceedings of the International Conference on Consumer Electronics*, LasVegas, USA, 11–14 Jan 2013 (IEEE). http://dx.doi.org/10.1109/ICCE.2013.6486825
25. A. Liotta, The cognitive net is coming. IEEE Spectr. **50**(8), 26–31 (IEEE, 2013)
26. J. Katz, *Clinical Audiology* (Williams & Wilkins, Philadelphia, 2002)
27. C. Owsley, R. Sekuler, D. Siemsen, Contrast sensitivity throughout adulthood. Vis. Res. **23**(7), 689–699 (1983)
28. S. Winkler, *Digital Video Quality* (Wiley, Chichester, 2005)
29. J. Yang, W. Makous, Implicit masking constrained by spatial inhomogeneities. Vis. Res. **37**(14), 1917–1927 (1997)
30. M. Rabbani, P. Jones, *Digital Image Compression Techniques* (SPIE-International Society for Optical Engineering, Bellingham, 1991)
31. S. Kanumuri, P. Cosman, A. Reibman, V. Vaishampayan, Modeling packet-loss visibility in MPEG-2 video. IEEE Trans. Multimed. **8**(2), 341–355 (2006)
32. A. Reibman, V. Vaishampayan, Quality monitoring for compressed video subjected to packet loss, in *Proceedings of 2003 International Conference on Multimedia and Expo, ICME'03*, vol. 1 (IEEE, 2003). pp. I–17
33. M. Wada, Selective recovery of video packet loss using error concealment. IEEE J. Sel. Areas Commun. **7**(5), 807–814 (1989)
34. G. Exarchakos, V. Menkovski, L. Druda, A. Liotta, Network analysis on Skype end-to-end video quality. Int. J. Pervasive Comput. Commun. **11**(1), (Emerald, 2015). http://www.emeraldinsight.com/doi/abs/10.1108/IJPCC-08-2014-0044

35. G. Exarchakos, L. Druda, V. Menkovski, P. Bellavista, A. Liotta, Skype resilience to high motion videos. Int. J. Wavelets, Multiresolut. Inf. Process. **11**(3) (World Scientific Publishing, 2013). http://dx.doi.org/10.1142/S021969131350029X

36. W. Tan, A. Zakhor, Real-time internet video using error resilient scalable compression and TCP-friendly transport protocol. IEEE Trans. Multimed. **1**(2), 172–186 (1999)

37. M. Zink, O. Künzel, J. Schmitt, R. Steinmetz, Subjective impression of variations in layer encoded videos. Quality Service-IWQoS **2003**, 155–155 (2003)

38. K. Tan, M. Ghanbari, D. Pearson, An objective measurement tool for MPEG video quality. Signal Process. **70**(3), 279–294 (1998)

39. R. Steinmetz, Human perception of jitter and media synchronization. IEEE J. Select. Areas Commun. **14**(1), 61–72 (1996)

40. L. Mued, B. Lines, S. Furnell, P. Reynolds, The effects of lip synchronization in IP conferencing, in *International Conference on Visual Information Engineering, VIE 2003*. (IET, 2003), pp. 210–213

41. J. Hawkins, S. Blakeslee, *On Intelligence* (St. Martin's Griffin, New York, 2005)

42. V. Menkovski, G. Exarchakos, A. Liotta, The value of relative quality in video delivery. J. Mob. Multimed. **7**(3), 151–162 (2011)

43. V. Menkovski, G. Exarchakos, A. Liotta, Tackling the sheer scale of subjective QoE, *Mobile Multimedia Communications* (Springer, Berlin, 2012), pp. 1–15

44. D.C. Mocanu, A. Liotta, A. Ricci, M. Torres Vega, G. Exarchakos, When does lower bitrate give higher quality in modern video services? in *Proceedings of the 2nd IEEE/IFIP International Workshop on Quality of Experience Centric Management*, Krakow, Poland, 9 May 2014 (IEEE). http://dx.doi.org/10.1109/NOMS.2014.6838400

45. A. Ortega, K. Ramchandran, Rate-distortion methods for image and video compression. IEEE Signal Process. Mag. **15**(6), 23–50 (1998)

46. M. Farias, M. Moore, J. Foley, S. Mitra, 17.2: detectability and annoyance of synthetic blocky and blurry video artifacts. SID Symp. Dig. Tech. Pap. **33**(1), 708–711 (Wiley Online Library, 2012)

47. R. Apteker, J. Fisher, V. Kisimov, H. Neishlos, Video acceptability and frame rate. IEEE MultiMed. **2**(3), 32–40 (1995)

48. J. Okyere-Benya, M. Aldiabat, V. Menkovski, G. Exarchakos, A. Liotta, Video quality degradation on IPTV networks, in *Proceedings of International Conference on Computing, Networking and Communications*, Maui, Hawaii, USA, Jan 30–Feb 2 2012 (IEEE)

49. C. Shannon, W. Weaver, R. Blahut, B. Hajek, *The Mathematical Theory of Communication*, vol. 117 (University of Illinois press, Urbana, 1949)

50. T. Berger, *Rate Distortion Theory: A Mathematical Basis for Data Compression* (Prentice-Hall, Englewood Cliffs, 1971)

51. L. Superiori, C. Weidmann, O. Nemethova, Error detection mechanisms for encoded video streams, in *Video and Multimedia Transmissions over Cellular Networks: Analysis, Modelling and Optimization in Live 3G Mobile Networks* (2009), p. 125

52. U. Horn, K. Stuhlmüller, M. Link, B. Girod, Robust internet video transmission based on scalable coding and unequal error protection. Signal Process. Image Commun. **15**(1), 77–94 (1999)

53. J. Chakareski, B. Girod, Rate-distortion optimized video streaming over internet packet traces, in *IEEE International Conference on Image Processing, 2005. ICIP 2005*, vol. 2. (IEEE, 2005), pp. II–161

54. M. Alhaisoni, A. Liotta, M. Ghanbari, Resource-awareness and trade-off optimization in P2P video streaming. Int. J. Adv. Med. Commun. (special issue on High-Quality Multimedia Streaming in P2P Environments) **4**(1), 59–77 (Inderscience Publishers, 2010). doi:10.1504/IJAMC.2010.030005, ISSN:1741-8003

55. K. Seshadrinathan, R. Soundararajan, A. Bovik, L. Cormack, A subjective study to evaluate video quality assessment algorithms, in *SPIE Proceedings Human Vision and Electronic Imaging*, vol. 7527 (Citeseer, 2010)

56. V. Menkovski, A. Oredope, A. Liotta, A. Cuadra-Sanchez, Optimized online learning for QoE prediction, in *Proceedings of the 21st Benelux Conference on Artificial Intelligence*, Eindhoven, The Netherlands, 29–30 Oct 2009, pp. 169–176. http://wwwis.win.tue.nl/bnaic2009/proc.html, ISSN:1568-7805

57. V. Menkovski, G. Exarchakos, A. Liotta, Online QoE Prediction, in *Proceedings of the 2nd IEEE International Workshop on Quality of Multimedia Experience*, Trondheim, Norway, 21–23 June 2010 (IEEE)

58. V. Menkovski, G. Exarchakos, A. Liotta, Online learning for quality of experience management, in *Proceedings of the Annual Machine Learning Conference of Belgium and The Netherlands*, Leuven, Belgium, 27–28 May 2010. http://dtai.cs.kuleuven.be/events/Benelearn2010/submissions/benelearn2010_submission_20.pdf

59. A. Liotta, Farewell to deterministic networks, in *Proceedings of the 19th IEEE Symposium on Communications and Vehicular Technology in the Benelux*, Eindhoven, The Netherlands, 16 Nov 2012 (IEEE). http://dx.doi.org/10.1109/SCVT.2012.6399413

60. G. Sullivan, T. Wiegand, Rate-distortion optimization for video compression. IEEE Signal Process. Mag. **15**(6), 74–90 (1998)

Chapter 2
Objective QoE Models

Abstract QoE is a metric that relates to our subjective expectations, and even though these expectations are not objectively measurable, many factors that contribute to them are. For example we can measure the loss of IP packets in the network and make an estimation on the effect that this will have on QoE. Similarly, we can measure the amount of signal degradation that a lossy compression process inflicts on the content. These measurements do not convey the exact difference between the expected and the delivered quality in a general case, but for more specific uses they can provide a good indication. The models that contain objectively collected measurements of factors that affect QoE are referred to as objective QoE models. This chapter discusses different objective QoE models and how they can be a part of a QoE management framework.

2.1 Objective Aspects of QoE

The main motivation for the use of the objective methods is that the objective factors can be measured precisely and at a lower cost than subjective assessment. Furthermore, many such methods can be deployed on a wide range of systems and their operation can be efficiently automated. Due to this, objective methods are frequently used for modelling the system quality and in-turn optimizing multimedia services.

Since video quality has such a significant importance on the overall QoE management of multimedia streaming services [1–16], this section is dedicated to a review of a range of methods for objective video quality assessment (VQA). VQA has been of significant importance since the early days of digital video, so many methods have been devised. The objective quality methods are further divided into three groups based on the level of involvement of the original reference signal in the estimation. The Full-reference (FR) methods require the original material in its entirety. They operate by comparing the original with the impaired material to calculate the degradation. The calculation ranges from simple algorithms, such as signal error estimates, to very complex ones that incorporate many HVS characteristics in the estimation. The Reduced-reference (RR) methods use parts or digests of the original material for the comparison calculation. They are better suited for situation where

© Springer International Publishing Switzerland 2015
V. Menkovski, *Computational Inference and Control of Quality in Multimedia Services*, Springer Theses, DOI 10.1007/978-3-319-24792-2_2

the original content is difficult to store or transport to the place of estimation or computational power is limited. Finally, No-reference (NR) methods do not use any part of the original content. They do not rely on comparisons but on measurements of external factors to model the QoE. The NR methods often are significantly restricted for specific applications and are not applicable for general use, but require the least resources and are useful for cases where the original content is not available.

As the goal of this work is to develop efficient methods for managing QoE of multimedia services, first we need to understand how to measure it. In light of this, we present an overview on the most commonly used objective QoE metrics as well as our experimental analysis of them in the rest of this chapter.

2.2 State of the Art

There are a vast number of objective video quality assessment methods [17, 18]. Some have evolved from image quality assessment, others have been particularly designed for video. They range in complexity from simple and easy to implement to very complex and computationally expensive. They also vary in performance, some are very restricted with limited correlation with the subjective QoE, and others are much better correlated. The rest of this section presents a set of representative objective metrics, from very simple with low correlation to very complex with high correlation with subjective QoE.

2.2.1 PSNR

Peak signal to noise ratio (PSNR) is one of the most commonly used FR VQA. The method is designed for a more general use, as it computes errors in any type of signal, and is also intensely used for image quality assessment (IQA) and VQA due to its simplicity.

PSNR estimates the difference between the original image and the distorted one by calculating the mean squared error (MSE) (Eq. 2.1) between the two signals and giving the ratio between the maximum of the signal and the MSE (Eq. 2.2), where x_{ij} is the value of the pixel in the original image at coordinates (i, j); y_{ij} is the value of the pixel at the same coordinates in the impaired image; and where MAX_I is the maximum amplitude of the pixel values in the image.

$$MSE = \frac{1}{mn} \sum_{i=0}^{m-1} \sum_{j=0}^{n-1} (x_{ij} - y_{ij})^2 \tag{2.1}$$

$$PSNR = 10\log_{10} \left(\frac{MAX_I^2}{MSE} \right)^2 \tag{2.2}$$

Regardless of its significant drawbacks (mainly its low correlation with subjective estimations) [19–23], PSNR is still very present in video quality analysis. It is easy to compute and provides a first impression on the quality achieved.

Different studies in VQA have shown that PSNR shows certain level of correlation to subjective quality when small number of factors is considered [24]. Typical example is the quantization level effect on quality during the compression of video. When all other factors are constant the effect of the quantization in the encoder tends to correlate well with PSNR. In fact, PSNR is also used for quality decisions in the encoder [25].

2.2.2 SSIM

The structural similarity index (SSIM) is a method that was originally developed for IQA [26], but is widely used for VQA as well. SSIM does not purely focus on bit-errors, but more on the changes in the structure of the image. In this way it addresses some of the drawbacks of PSNR, such as susceptibility to changes in brightness and contrast. The HVS demonstrates luminance and contrast masking, which SSIM takes into account while PSNR does not. The HVS demonstrates light adaptation characteristics and as a consequence of that it is sensitive to relative changes in brightness. This effect is referred to as luminance masking. On the other hand, changes in contrast are less noticeable when the base contrast is high than when it is low. This effect is referred to as contrast masking.

The SSIM index executes three comparisons, in terms of luminance, contrast and structure. The output value is a combination of these three comparisons as given in Eq. 2.3, where x and y are vectors containing the pixel values of the original and the impaired image, respectively.

$$SSIM(x, y) = f(l(x, y), c(x, y), s(x, y)) \tag{2.3}$$

For the luminance comparison, first the mean luminance is calculated (Eq. 2.4).

$$\mu_x = \frac{1}{N} \sum_{i=1}^{N} x_i \tag{2.4}$$

Then a luminance comparison is executed as in Eq. 2.5.

$$l(x, y) = \frac{2\mu_x \mu_y + C_1}{\mu_x^2 + \mu_y^2 + C_1} \tag{2.5}$$

The value of the parameters $C1$ is set to $(K_1 L)^2$, where $K_1 \ll 1$ is a small constant and L is the dynamic range of the pixel values. The average luminance is removed

from the signal amplitude and a contrast comparison is computed. The base contrast
of each signal is computed using its standard deviation (Eq. 2.6).

$$\sigma_x = \left(\frac{1}{N-1} \sum_{i=1}^{N} (x_i - \mu_x)^2 \right)^{\frac{1}{2}} \tag{2.6}$$

The contrast comparison is computed as given in Eq. 2.7.

$$c(x, y) = \frac{2\sigma_x \sigma_y + C_2}{\sigma_x^2 + \sigma_y^2 + C_2} \tag{2.7}$$

For the structure comparison, the average luminance is subtracted and is divided
by its base contrast to normalize it. A Pearson correlation coefficient is calculated as
a measure of structural similarity (Eqs. 2.8 and 2.9).

$$s(x, y) = \frac{\sigma_{xy} + C_3}{\sigma_x \sigma_y + C_3} \tag{2.8}$$

$$\sigma_{xy} = \frac{1}{N-1} \sum_{i=1}^{N} (x_i - \mu_x)(y_i - \mu_y) \tag{2.9}$$

The SSIM output is a combination of all three components (Eq. 2.10).

$$SSIM(x, y) = [l(x, y]^\alpha [c(x, y)]^\beta [s(x, y)]^\gamma \tag{2.10}$$

This SSIM model is parameterized by α, β, γ where typically the parameter
values are α, β and $\gamma = 1$. In order to simplify the expression the parameter C_3 is set
to $C_3 = C_2/2$ (Eq. 2.11).

$$SSIM(x, y) = \frac{(2\mu_x \mu_y + C_1)(2\sigma_{xy} + C_2)}{(\mu_x^2 + \mu_y^2 + C_1)(\sigma_x^2 + \sigma_y^2 + C_2} \tag{2.11}$$

Since the luminance calculation (2.6) and the contrast calculation (2.7) are consistent with the luminance and contrast masking effects respectively, the SSIM metric
performance is better correlated with subjective QoE [23].

2.2.3 VQM

Methods developed for IQA can deliver some level of indication as to the degradation
of video quality. However, these methods are not designed for video and omit the
temporal factors' effects on the quality. One such example is the motion masking

effects, where high motion in the video decreases the effect of loss of structure on the quality [27]. A model that addresses the quality of video, taking into account the structural and temporal aspects, is the video quality model (VQM) [28].

Because of its good correlation with subjective values, VQM is commonly used as a FR method. However, VQM does not compare the original to the impaired video directly. It extracts features from the original and the impaired video separately, and then compares those features to calculate quality. This makes the method applicable for RR use as well. The features that are extracted from the original video account for 9.3 % of the uncompressed size of the video. Additionally another 4.5 % of data needs to be transmitted for the initial pre-processing step of VQM where both videos are calibrated in space and in time.

VQM is implemented by first applying a perceptual filter to the video stream. This enhances some properties of perceived video quality, such as the edge information. After this perceptual filtering, the video is segmented in space and time into spatial-temporal (S-T) subregions. Next the features are extracted from these S-T subregions. Finally, a perceptibility threshold is applied to the extracted features, so that only impairments above this threshold are considered.

The masking effects in the HVS imply that impairment perception is inversely proportional to the amount of localized spatial or temporal activity that is present. In other words, spatial impairments become less visible as the spatial activity increases, and temporal impairments become less visible as the temporal activity increases. Furthermore, these masking effects interact with each other, so spatial masking has effect on temporal quality perception and vice versa. To account for these effects, in VQM the perceptual impairment at each S-T region is calculated using comparison functions. Some features use a comparison function that performs a simple Euclidean distance between two original and two processed feature streams. But more commonly, features use either the ratio comparison function or the log comparison function.

After different impairment parameters have been calculated in different spatial and temporal regions, these values need to be collapsed into a single value for the quality index. Optimal spatial collapsing function often involves some form of worst case processing, such as taking the average of the worst 5 % of the distortions observed at a particular point in time. Because localized impairments tend to draw the focus of the viewer, making the worst part of the picture the predominant factor in the subjective quality decision is a good strategy. Finally spatial and temporal collapsing functions are used to produce a single objective quality value for the video sequence.

The parameters included in the VQM model are the following:

- si_loss: detects a decrease or loss of spatial information (e.g., blurring);
- hv_loss: detects a shift of edges from horizontal and vertical orientation to diagonal orientation; this might be the case if horizontal and vertical edges suffer more from blurring than diagonal edges;
- hv_gain: detects a shift of edges from diagonal to horizontal and vertical; this might be the case if the processed video contains tiling or blocking artefacts;

- chroma_spread: detects changes in the spread of the distribution of two-dimensional colour samples;
- si_gain: measures improvements to quality that result from edge sharpening enhancements;
- ct_ati_gain: accounts for the interactions between the amount of spatial detail and motion on the perceived of spatial and temporal impairments (spatial and temporal masking);
- chroma_extreme: detects severe localized colour impairments, such as those produced by digital transmission errors.

The General Model is a weighted linear combination of these parameters (Eq. 2.12). The weights given in Eq. 2.12 are selected to achieve maximum objective to subjective correlation for a wide range of video quality and bit rates.

$$\begin{aligned}
VQM = &- 0.2097 * si\_loss \\
&+ 0.5969 * hv\_loss \\
&+ 0.2483 * hv\_gain \\
&+ 0.0192 * chroma\_spread \\
&- 2.4316 * si\_gain \\
&+ 0.0431 * ct\_ati\_gain \\
&+ 0.0076 * chroma\_extreme \qquad (2.12)
\end{aligned}$$

2.2.4 MOVIE

MOtion-based Video Integrity Evaluation (MOVIE) is another VQA index that integrates both spatial and temporal aspects [29]. It implements a spatio-temporally localized, multi-scale decomposition of the reference and test videos using a set of spatio-temporal Gabor filters [30]. The MOVIE index is composed of two components. The first one is the spatial MOVIE index, which uses the output of the multi-scale decomposition of the reference and test videos to measure spatial distortions in the video. The second one is the temporal MOVIE index, which captures temporal degradations. The Temporal MOVIE index first computes the motion information from the reference video to generate motion trajectories. Then it evaluates the temporal quality of the test video along the computed motion trajectories of the reference video. In this way MOVIE attempts to account for the motion processing of the HVS and capture the intensity of the temporal distortions as would be perceived by the viewer.

Both MOVIE components work together. The spatial quality map generated by the spatial MOVIE, responds to the blur in the test video. The temporal quality generated by the temporal MOVIE, maps motion compensation mismatches on the edges.

The maps are then collapsed into two indexes. This is done by calculating the ratio of the standard deviation to the mean of the values in the map. This statistics is known as a coefficient of variation, and is a good predictor of the subjective quality of a video.

Even though, MOVIE is an objective method that correlates well with subjective feedback from viewers, its high cost in computational power and memory limits its implementation in real-time systems.

2.2.5 Reduced and No Reference Methods

A more suitable alternative for real-time quality estimation in content delivery systems is given by the RR and NR methods. These methods are also applicable when the original content is not available, e.g. when different video filtering is applied (denoiseing, deinterlacing, resolution upscaling or due to storage restrictions). These methods, or models, are usually much more restricted than the FR methods. Particularly the NR methods, only deal with specific types of impairments and are not very accurate for general use.

Gunawan and Ghanbari have developed a RR method that uses local harmonic strength features from the original video to calculate the amount of impairments or quality of the affected video [31]. Harmonic gains and loss correlate well with two very common types of impairment present in MPEG encoded video, i.e. blockiness and blurriness. The features (harmonic data) in this RR method have a very low overhead of only 160–400 bits/s, which is a negligible amount compared to the size of the video.

Ma et al. present a RR method that generates both spatial and temporal features [32]. From the spatial perspective they define an energy variant descriptor (EVD) to measure the energy change in each individual encoded frame, which results from the quantization process. The EVD is calculated as the proportion of the medium plus high frequencies in the image to low frequencies. The EVD of the original video is then compared to the EVD of the compressed video. Due to the fact that different frequencies are quantized with different fidelity in a lossy compression process, the EVD difference will indicate a level of impairment. The temporal features are collected from the difference between two adjacent frames. On these computed difference frames, a generalized Gaussian density function is employed to extract these features. Then a city block distance is used to calculate the distance between the features of the original video and the impaired one. Finally all of the distances are combined to produce the quality index. The authors claim that even though this is a RR method it outperforms simpler FR methods such as PSNR and SSIM.

NR methods are even more flexible than RR because they are applicable to any video environment, even ones that do not have any information on the original source of the video. Naturally their accuracy and generality is highly constrained. NR models are frequently used to calculate the impact of transport errors on the delivered video. In [33] authors present a model for estimating MSE caused by packet loss, by examining the video bit-stream. A single packet loss does not affect only the pixels of the video frame that lost information in that packet but, due to the temporal compression mechanisms of MPEG videos, these errors are propagated by the motion vectors in the subsequent frames. The location of the packet is of significant importance, because different types of frames carry information of different nature. A similar approach for MPEG2 video is presented in [34]. This method uses different machine learning (ML) algorithms to predict the visibility of the lost packet on the presented video. The additional complexity that these NR methods face is that they are not aware of the decoder's approach to conceal the error. A typical concealment approach is zero-motion concealment, in which a lost macro-block is concealed by a macro-block in the same location from a previous frame. However, the visibility of this concealment depends on the content at this position, size of the screen and many other factors that are generally related to the overall QoE.

In more recent developments, different researchers have attempted to use RR and NR methods to assess QoE in real-time, that is during a video streaming session (and without delaying it) [35]. Mocanu et al. have been working on deep learning techniques for both 2D and 3D images [36, 37]. Torres Vega et al. have been using cognitive methods to assess streaming videos on lightweight devices [38, 39], addressing the limitation of earlier work based on FR metrics [40–46].

2.3 Models for JPEG2000 and MPEG4/AVC

During the encoding process of multimedia content the encoder continually makes decision that the delivered bit-rate as well as the fidelity of the compressed signal. Many optimization techniques are developed to help make the most efficient decisions. These optimization techniques commonly rely on objective quality models to evaluate the effect of different decisions. In this section we present a discussion on objective models used in the encoding process of images in JPEG2000 [47] and for video in MPEG4/AVC [48].

Commonly JPEG2000 implementations adopt the rate-based distortion minimization encoding approach that requires the user to specify the desired bit-rate or the desired quality level. The encoder then needs to match the desired bit-rate, while minimizing the loss of quality in the image or provide a standardized level of quality for the minimum bit-rate. For both of these modes the encoders need to understand the relationship between the bit-rate and the quality.

The simplest models for distortion are based on the MSE calculation (e.g. PSNR). Even though these objective metrics do not correlate well with the perception of the HVS their simplicity makes them an attractive option for many encoder implementations. However, in other implementations some characteristics of the HVS are used to improve the accuracy of the MSE based models. The basic idea is to remove perceptually irrelevant information, that is, information when removed introduce artifacts that are imperceptible or bellow the threshold of detection by the receiver. In the bit-rate mode that would mean eliminating all artifacts up to the point that produces the desired size of image, while ordering the artifacts based on their how sensitive the viewers is to them. And in the constant quality mode, eliminating all the artifacts that create a distortion above the desired level.

One of the better understood aspects of HVS is the contrast sensitivity [49]. The contrast sensitivity function (CSF) describes the sensitivity of the human eye to different spatial frequencies. Instead of using MSE a CSF weighted MSE function is introduced, which more efficiently reports the quality [49]. Artifacts that are superimposed on a non uniform background with similar spatial frequency are much less noticeable. This type of masking effect is exploited to improve the efficiency of RDO. Furthermore, an improvement in encoding efficiency in JPEG2000 is introduced by adjusting the quantization step for each spatial frequency band individually and in accordance with the CSF. In other words, the degradation will be calculated higher per bit for spatial frequencies, for which the human eye is more sensitive to.

The perceptual distortion coding presented in [50] attempts to discriminate between signal components which are detected and are not detected by the HVS. The models approach is to 'hide' the coding distortion beneath the detection threshold and to remove perceptually irrelevant signal information. In this method three visual phenomenon are used to calculate the thresholds: contrast sensitivity, luminance masking and contrast masking. The thresholds are defined by the smallest contrast that yields a visible signal over background of uniform intensity. Constant a key factor because the HVS perception depends much less on the absolute luminance perceived, but rather on the variation in the signal relative to its surrounding background. This phenomena is known as Weber-Fechner's law [51]. After calculating the localized thresholds the method uses a probability spectral and spatial summation model to develop an overall perceptual distortion metric. The method is suitable for generating consistent quality images at lower bit-rates.

A particular type of encoding, referred to as Embedded coding [52], is designed for sending images over a network. The images encoded with Embedded encoding if truncated are decoded into a visible image. This type of images can also be decoded progressively, improving the overall experience of the user. The image is encoded from the most significant bit-plane to the least significant bit-plane. A RDO strategy adapted to the embedded approach is the visual progressive weighting approach. In this approach much more aggressive weighting strategy is implemented in the beginning of the bit-stream with the more significant bit-planes and a less aggressive as decoding proceeds and quality improves [53].

In video coding the RDO is further complicated by the temporal component. The temporal aspect introduces additional masking effects that can be leveraged for reduction of bit-rate. However, in video coding the complexity of maintaining constant level of quality over the different frames is is much higher [48].

One of the most commonly used video coding standards now is H.264 (MPEG-4 Part 10). H.264 as other standards before (H.263 and MPEG-2) uses translational block-based motion compensation and transform based residual coding [48]. The output bit-rate can be controlled by several coding parameters the quantization scale and the coding mode. Large quantization scale reduces the bit-rate, but also the fidelity of the compressed video. The RDO problem is typically divided into three subproblems: Group of pictures (GOP) bit allocation; frame bit allocation; and macroblock quantization parameter (Q) selection. For CBR allocation, first the GOP is allocated the selected amount of bits, than this amount is distributed to all the frames in the GOP. The distribution is made in such manner that the quality of the frames is kept as constant as possible. Finally, a similar approach is taken to redistribute the allotted bits within the frame to the macroblocks [54]. Measuring the quality or the distortion is commonly done using PSNR of the compressed against the original signal.

The specification of H.264 does not define the way that the encoder is implementing the RDO, this is left to the developers. The reference H.264 software uses RDO in which the Lagrangian multiplier λ of the cost function $J = D + \lambda R$ is selected taking into account the quantization value. An optimal solution for this optimization would require perfect understanding of the characteristics of the video content and the effects on the encoder parameters on it. Since this understanding is not available, the parameters are either selected by executing several encoding passes and observing the results [55] or using models for the rate-distortion effects based on different parameter values [56]. The later approach is more favorable in many approaches where the encoding process is time sensitive.

In the transform based coding approaches the macroblock data is transformed into a set of coefficients using DCT (Discrete cosine transform). The coefficients are than quantized and encoded with a variable-lenght coding [56]. Due to this number of bits and the distortion for a given macroblock depend on the quantization parameter and the entropy of the coefficients.

$$R(Q) \approx H(Q) \tag{2.13}$$

where $H(Q)$ is the entropy of the DCT coefficients. The entoropy of the coefficients is typically described by a Laplacian distribution [56] and the rate function is modeled as given in (2.14).

$$R(Q) = \begin{cases} \frac{1}{2}\log_2(2e^2\frac{\sigma^2}{Q^2}), & \frac{\sigma^2}{Q^2} > \frac{1}{2e} \\ \frac{e}{\ln 2}\frac{\sigma^2}{Q^2}, & \frac{\sigma^2}{Q^2} \leq \frac{1}{2e} \end{cases} \tag{2.14}$$

The distortion in the ith macroblock is introduced by uniformly quantizing the DCT coefficients with a step size Q_i. This model defines the distortion model as given in (2.15).

$$D = \frac{1}{N}\sum_{i=1}^{N}\frac{Q_i^2}{12} \tag{2.15}$$

In another model He et al. present the R-D model based on the fractions of zeros among the quantized DCT coefficients ρ [57]. This model also make the assumption that the coefficients are distributed with a Laplacian distribution. The model represents the rate with a linear dependency from ρ (Eq. 2.16).

$$R = \theta(1 - \rho) \tag{2.16}$$

and the distrtion as given in Eq. 2.17.

$$D = \sigma^2 e^{-\alpha(1-\rho)} \tag{2.17}$$

where θ and α are model parameters. The authors claim that this model improves the delivered quality by an average 0.3 dB [58].

Even though, most commonly generalized Gaussian or Laplacian distribution is assumed for the DCT coefficients, methods exist that use other distributions have also been proposed (Cauchy [54]), however more complex distribution lead to increased complexity in the models.

RDO optimization offers a key powerful mechanisms for optimizing the trade-off between the bit-rate and quality. However, these sophisticated the optimization models still rely on the simple objective quality (distortion) measures due to the strict restrictions in computational complexity in the domain of multimedia encoding.

2.3.1 Characterizing the Video Content

The perceived QoE depends on the type of video content as well [12, 59–61]. In order to build a more comprehensive QoE model we need to incorporate information about the video content. Defining features that can be automatically extracted from the video and that will carry information about the type of content is not straight forward.

Two features that are expected to correlate well with the difficulty in compressing videos are the Spatial Information (SI) and Temporal information (TI) indexes. These can also to a certain extent correlate with the type of the content [62–64].

The SI index carries the amount of spatial information in each frame of the video. The more structural details or edges are present in the video the higher the SI index will be. The SI is calculated as the standard deviation over both the x and y directions of the frame (spatial standard deviation) after the frame data has been put through a Sobel filter [65].

$$SI[F_n] = STD_{space}[Sobel(F_n)] \tag{2.18}$$

The *Sobel* function given in Eq. 2.18 implements the Sobel filter, which is used to extract the edge structure information from the image [66]. The F_n variable refers to the nth frame of the video.

The TI index carries information about the amount of temporal information in the video, or the intensity of changes in the video over time. This is proportional to the amount of movement in the video.

The TI is calculated as the difference between two frames and a spatial standard deviation on that difference (Eqs. 2.19 and 2.20).

$$\Delta F_n = F_n - F_{n-1} \tag{2.19}$$

$$TI[F_n] = STD_{space}[\Delta F_n] \tag{2.20}$$

Videos with similar SI, TI values tend to contain similar type of content or have similar characteristics. For example 'head and shoulders' videos, which are common for news broadcast, tend to have low SI and TI. On the other, hand videos with complex scenes and high amount of movement such as action movies will have high SI and TI. In a similar fashion, different types of content such as football matches, documentaries or music videos tend to make separate clusters of SI and TI combinations. For this reason SI and TI can be useful features to convey the type of content for the QoE modelling.

Qualitatively similar types of features can be collected as a by product from the encoding process itself [67]. In this case no additional calculation is necessary to obtain these features. Hu and Wildfeuer define two indexes, one for scene complexity C and one for level of motion M that can be computed based on the amount of data in the I frames and the P frames of the encoded video.

With a typical Group of Pictures type encoding (such as the H.264) the amount of data in the I-frame corresponds to the level of complexity in the image [67] (Eq. 2.21). This is due to the fact that the more low frequency components there are in the image the more bits the encoder needs to use in the I-frame to encode the frame if the encoding is set to a constant quality mode. Similarly the P-frame corresponds to the amount of changes that happened since the last I-frame, so it correlates well with the amount of movement in the video (Eq. 2.22).

$$C = \frac{B_I}{2 \cdot 10^6 \cdot 0.91^{QP_I}} \tag{2.21}$$

$$M = \frac{B_P}{2 \cdot 10^6 \cdot 0.87^{QP_P}} \tag{2.22}$$

In Eqs. 2.21 and 2.22, B_I and B_P correspond to the number of bytes in the I-frames and P-frames respectively and QP_I and QP_P is the quantization parameter for the I-frames and the P-frames (measured across the whole frame-set of the coded sequences).

2.4 Experimental Analysis of Objective QoE

VQA indexes vary in complexity and how in accurately they correlate with subjective estimations. However, in specifically constrained conditions their evaluations can bring valuable information for the delivered quality. Understanding these constraints can deliver a useful tool for a more general QoE model.

Table 2.1   Description of the live videos

bs	Blue sky	Circular camera motion showing a blue sky and some trees
rb	River bed	Still camera, shows a river bed containing some pebbles in the water
pa	Pedestrian area	Still camera, shows some people walking about in a street intersection
tr	Tractor	Camera pan shows a tractor moving across some fields
sf	Sunflower	Still camera, shows a bee moving over a sun-flower in close-up
rh	Rush hour	Still camera, shows rush hour traffic on a street
st	Station	Still camera, shows railway track, a train and some people walking across the track
sh	Shields	Camera pans at first, then becomes still and zooms in; shows a person walking across a display pointing at it
mc	Mobile and Calendar	Camera pan, tor train moving horizontally with a calendar moving vertically in the background
pr	Park run	Camera pan, a person running across a park

To explore this, we implement an experiment where the content is impaired only with lossy compression. The level of impairment is controlled with a constant bit-rate (CBR) level of encoding. In this setting we observe how the output of different objective VQA metrics changes with the type of content.

The raw video samples we used for this assessment are part of the Live Video Quality Database [68, 69]. The description of each video is given in Table 2.1. Snaphots of four videos from the database are given in Fig. 2.1.

Fig. 2.1   Snapshots from 4 of the used videos. Starting from *top left* image in *clockwise order*: river bed, park run, sunflower, mobile and calendar

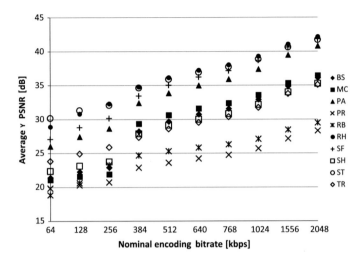

Fig. 2.2 PSNR calculated quality degradation from CBR compression over different bit-rates

The ten different videos were compressed with H.264 compression [70]. The PSNR index was calculated against the original uncompressed sequences. The videos native resolution is 768×432 at 25 frames/s. The videos were compressed with bit-rate settings ranging from 64 kb/s to 2 Mb/s. To achieve the very low bit-rate of 64 kb/s, the video had to be spatially and temporally sub-sampled to 384×216, at 12.5 frames/s.

2.4.1 PSNR Results

The PSNR calculations were executed (frame-by-frame) as defined in Eq. 2.2, on the luminance (Y) component of the raw original sequence and the sequence impaired with compression. The mean values for the calculations over all the frames in the figure are given in Fig. 2.2.

2.4.2 SSIM Results

The SSIM calculations were executed in a similar manner as the PSNR. Videos are impaired with different level of compression and on each pair of frames (original and impaired) SSIM is calculated. The mean values for each sequence are given in Fig. 2.3.

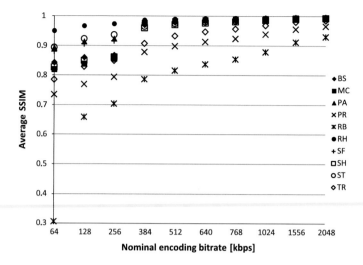

Fig. 2.3 SSIM calculated quality degradation from CBR compression over different bit-rates

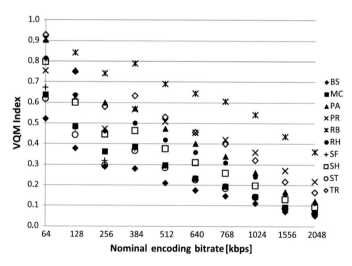

Fig. 2.4 VQM calculated quality degradation from CBR compression over different bit-rates

2.4.3 VQM Results

The VQM calculations were implemented using the VQM reference software [71]. The VQM was executed again on pairs of original and impaired videos, for each level of impairment. The results of the VQM calculations are given in Fig. 2.4.

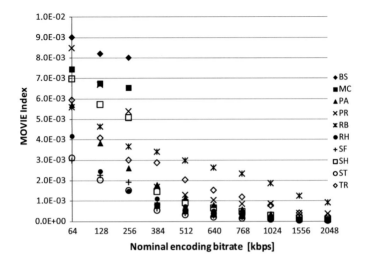

Fig. 2.5 MOVIE calculated quality degradation from CBR compression over different bit-rates

2.4.4 MOVIE Results

Finally we executed the MOVIE index on the database of videos we developed for this experiment. The MOVIE reference software was received courtesy of the authors [72]. The results are shown in Fig. 2.5.

2.4.5 SI, TI, C and M Results

The description of each type of video can be obtained from the Live database. However, in order to objectively generalize on the types of video we set out to compute the spatial and temporal features of the videos. In this way we can analyze if there is any correlation between the spatial and temporal features of the video and the VQA indexes output. The results of the SI, TI, C and M estimation are given in Fig. 2.6 and Fig. 2.7 respectively.

2.5 Discussion and Conclusions

The results from the objective VQA show that the PSNR index seems linearly proportional with the selected rate of bit-rate increase (Fig. 2.2). In contrast, SSIM shows more nonlinear drop in estimations (Fig. 2.3). This is more akin to what would be expected from a subjective evaluation. It is perceptually evident that the videos in

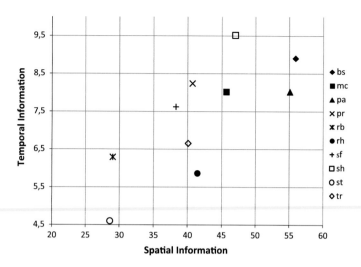

Fig. 2.6 Spatial and temporal information of the videos part of the objective VQA

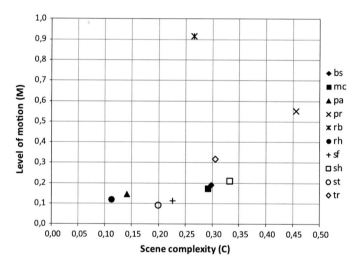

Fig. 2.7 Scene complexity and level of motion in the videos part of the objective VQA

the lower bit-rate range are significantly more degraded than in the higher range. The VQM results present an inversely proportional quality index to the level of bit-rate. But the values do not demonstrate the level of nonlinearity that one would expect for this set of bit-rate levels (Fig. 2.4). Finally, the MOVIE index demonstrates a clearly emphasized non-linear response (Fig. 2.5). Above 256 kb/s the quality index drops to much lower values (lower is better). This indicates that MOVIE finds the improvement in quality with much higher gradient in the lower bit-rate region than in the higher region. These results are better correlated with the measured subjective perception of quality both by the authors [69] and our own measurements discussed in the following chapter [73].

The evaluation of the SI and TI does not show a clear representation why certain videos are compressed with less quality than others using the same bit-rate. However, there are some indications. The videos that have both low SI and high TI demonstrate worse VQA indexes than the others (Fig. 2.6). On the other hand from the results of the 'scene complexity' (C) and 'level of motion' (M) indexes (Fig. 2.7) the relationship to the performances is much more evident. Clearly the worse performing videos 'river-bed' (rb) and 'park run' (pr) stand out with exaggeratedly higher C and M values. This high correlation is to be expected since the C and M values are derived from the encoder directly. Nevertheless, this presents a good indication of the quality of these features for characterizing the type of content in QoE models.

References

1. S. Winkler, P. Mohandas, The evolution of video quality measurement: from PSNR to hybrid metrics. IEEE Trans. Broadcast. **54**(3), 660–668 (2008). http://ieeexplore.ieee.org/xpls/abs_all.jsp?arnumber=4550731
2. G. Exarchakos, V. Menkovski, L. Druda, A. Liotta, Network analysis on Skype end-to-end video quality. Int. J. Pervasive Comput. Commun. **11**(1) (Emerald, 2015). http://www.emeraldinsight.com/doi/abs/10.1108/IJPCC-08-2014-0044
3. G. Exarchakos, L. Druda, V. Menkovski, P. Bellavista, A. Liotta, Skype resilience to high motion videos. Int. J. Wavelets, Multiresolut. Inf. Process. **11**(3) (2013). http://dx.doi.org/10.1142/S021969131350029X
4. F. Agboma, A. Liotta, Quality of Experience Management in Mobile Content Delivery Systems, J. Telecommun. Syst. (special issue on the Quality of Experience issues in Multimedia Provision) **49**(1), 85–98 (2012). doi:10.1007/s11235-010-9355-6
5. M. Alhaisoni, A. Liotta, M. Ghanbari, Scalable P2P video streaming. Int. J. Bus. Data Commun. Netw. **6**(3), 49-65 (2010). doi:10.4018/jbdcn.2010070103, ISSN:1548-0631
6. V. Menkovski, G. Exarchakos, A. Liotta, A. Cuadra Sánchez, Quality of experience models for multimedia streaming, Int. J. Mob. Comput. Multimedia Commun. **2**(4), 1–20 (IGI Global, Oct–Dec 2010). www.igi-global.com/ijmcmc/, doi:10.4018/jmcmc.2010100101, ISSN:1937-9412
7. A. Liotta, L. Druda, G. Exarchakos, V. Menkovski, Quality of experience management for video streams: the case of Skype, in *Proceedings of the 10th International Conference on Advances in Mobile Computing and Multimedia*, Bali, Indonesia, 3–5 Dec 2012 (ACM). doi:http://dx.doi.org/10.1145/2428955.2428977
8. J. Okyere-Benya, M. Aldiabat, V. Menkovski, G. Exarchakos, A. Liotta, Video quality degradation on IPTV networks, in *Proceedings of International Conference on Computing, Networking and Communications*, Maui, Hawaii, USA, 30 Jan–2 Feb 2012 (IEEE)
9. V. Menkovski, G. Exarchakos, A. Liotta, Online learning for quality of experience management, in *Proceedings of The annual machine learning conference of Belgium and The Netherlands*, Leuven, Belgium, 27th–28th May 2010. http://dtai.cs.kuleuven.be/events/Benelearn2010/submissions/benelearn2010_submission_20.pdf
10. V. Menkovski, G. Exarchakos, A. Cuadra-Sanchez, A. Liotta, Measuring quality of experience on a commercial mobile TV platform, in *Proceedings of the 2nd International Conference on Advances in Multimedia*, Athens, Greece, 13–19 June 2010 (IEEE)

11. V. Menkovski, G. Exarchakos, A. Liotta, Online QoE prediction, in *Proceedings of the 2nd IEEE International Workshop on Quality of Multimedia Experience*, Trondheim, Norway, 21–23 June 2010 (IEEE)

12. V. Menkovski, A. Oredope, A. Liotta, A. Cuadra-Sanchez, Predicting quality of experience in multimedia streaming, in *Proceedings of the 7th International Conference on Advances in Mobile Computing and Multimedia*, Kuala Lumpur, Malaysia, 14–16th Dec 2009 (ACM). ISBN:978-1-60558-659-5, http://dl.acm.org/citation.cfm?id=1821766

13. F. Agboma, M. Smy, A. Liotta, QoE analysis of a peer-to-peer television system, in *Proceedings of the International Conference on Telecommunications, Networks and Systems*. Amsterdam, Netherlands, 22–24 July 2008

14. F. Agboma, A. Liotta, QoE-aware QoS Management, in *Proceedings of the 6th International Conference on Advances in Mobile Computing and Multimedia*. Linz, Austria, 24–26 Nov 2008

15. F. Agboma, A. Liotta, Managing the user's quality of experience, in *Proceedings of the second IEEE/IFIP International Workshop on Business-driven IT Management (BDIM 2007)*, Munich, Germany, 21th May 2007 (IEEE)

16. F. Agboma, A. Liotta, User-centric assessment of mobile content delivery, in *Proceedings of the 4th International Conference on Advances in Mobile Computing and Multimedia*, Yogyakarta, Indonesia, 4–6 Dec 2006

17. S. Chikkerur, V. Sundaram, M. Reisslein, L. Karam, Objective video quality assessment methods: a classification, review, and performance comparison. IEEE Trans. Broadcast. **57**(2), 165–182 (2011)

18. V. Menkovski, A. Liotta, QoE for mobile streaming, in *Mobile Multimedia—User and Technology Perspectives*, ed. by D. Tjondronegoro (InTech, 2012). http://www.intechopen.com/books/mobile-multimedia-user-and-technology-perspectives/qoe-for-mobile-streaming

19. P. Teo, D. Heeger, Perceptual image distortion, in *Proceedings of IEEE International Conference on Image Processing. ICIP-94*, vol. 2 (IEEE, 1994), pp. 982–986

20. M. Eckert, A. Bradley, Perceptual quality metrics applied to still image compression. Signal Process. **70**(3), 177–200 (1998)

21. A. Eskicioglu, P. Fisher, Image quality measures and their performance. IEEE Trans. Commun. **43**(12), 2959–2965 (1995)

22. B. Girod, *What's Wrong with Mean-Squared Error?Digital Images and Human Vision* (MIT Press, Cambridge, 1993), pp. 207–220

23. Z. Wang, A. Bovik, L. Lu, Why is image quality assessment so difficult? in *IEEE International Conference on Acoustics, Speech, and Signal Processing (ICASSP), 2002*, vol. 4 (IEEE, 2002), pp. IV–3313

24. Q. Huynh-Thu, M. Ghanbari, Scope of validity of psnr in image/video quality assessment. Electron. Lett. **44**(13), 800–801 (2008)

25. F. Pan, X. Lin, S. Rahardja, K. Lim, Z. Li, D. Wu, S. Wu, Fast mode decision algorithm for intraprediction in h. 264/avc video coding. IEEE Trans. on Circuits Syst. Video Technol. **15**(7), 813–822 (2005)

26. Z. Wang, A. Bovik, H. Sheikh, E. Simoncelli, Image quality assessment: from error visibility to structural similarity. IEEE Trans. Image Process. **13**(4), 600–612 (2004)

27. Z. Wang, L. Lu, A. Bovik, Video quality assessment based on structural distortion measurement. Signal Process. Image Commun. **19**(2), 121–132 (2004)

28. M. Pinson, S. Wolf, A new standardized method for objectively measuring video quality. IEEE Trans. Broadcast. **50**(3), 312–322 (2004)

29. K. Seshadrinathan, A. Bovik, Motion-based perceptual quality assessment of video, in *IS&T/SPIE Electronic Imaging* (International Society for Optics and Photonics, 2009), pp. 72 400X–72 400X

30. R. Mehrotra, K. Namuduri, N. Ranganathan, Gabor filter-based edge detection. Patt. Recogn. **25**(12), 1479–1494 (1992)

31. I. Gunawan, M. Ghanbari, Reduced-reference video quality assessment using discriminative local harmonic strength with motion consideration. IEEE Trans. Circuits Syst. Video Technol. **18**(1), 71–83 (2008)

32. L. Ma, S. Li, K. Ngan, *Reduced-Reference Video Quality Assessment of Compressed Video Sequences* (2012)
33. A. Reibman, V. Vaishampayan, Quality monitoring for compressed video subjected to packet loss, in *Proceedings of International Conference on Multimedia and Expo, 2003. ICME'03. 2003*, vol. 1 (IEEE, 2003) pp. I–17
34. S. Kanumuri, P. Cosman, A. Reibman, V. Vaishampayan, Modeling packet-loss visibility in mpeg-2 video. IEEE Trans. Multimedia **8**(2), 341–355 (2006)
35. A. Liotta, D. Constantin Mocanu, V. Menkovski, L. Cagnetta, G. Exarchakos, Instantaneous video quality assessment for lightweight devices, in *Proceedings of the 11th International Conference on Advances in Mobile Computing and Multimedia*, Vienna, Austria, 2–4 Dec 2013 (ACM). http://dx.doi.org/10.1145/2536853.2536903
36. D.C. Mocanu, G. Exarchakos, H.B. Ammar, A. Liotta, Reduced reference image quality assessment via boltzmann machines, in *Proceedings of the 3rd IEEE/IFIP IM 2015 International Workshop on Quality of Experience Centric Management*, Ottawa, Canada, 11–15 May 2015 (IEEE)
37. D.C. Mocanu, G. Exarchakos, A. Liotta, Deep learning for objective quality assessment of 3D images, in *Proceedings of IEEE International Conference on Image Processing*, Paris, France, 27–30 Oct 2014 (IEEE)
38. M. Torres Vega, E. Giordano, D. C. Mocanu, D. Tjondronegoro, A. Liotta, Cognitive no-reference video quality assessment for mobile streaming services, in *Proceedings of the 7th International Workshop on Quality of Multimedia Experience*, Messinia, Greece, 26–29 May 2015 (IEEE) (http://www.qomex.org)
39. M. Torres Vega, D. Constantin Mocanu, R. Barresi, G. Fortino, A. Liotta, Cognitive streaming on android devices, in *Proceedings of the 1st IEEE/IFIP IM 2015 International Workshop on Cognitive Network & Service Management*, Ottawa, Canada, 11–15 May 2015 (IEEE). http://www.cogman.org
40. D.C. Mocanu, A. Liotta, A. Ricci, M. Torres Vega, G. Exarchakos, When does lower bitrate give higher quality in modern video services?, in *Proceedings of the 2nd IEEE/IFIP International Workshop on Quality of Experience Centric Management*, Krakow, Poland, 9 May 2014 (IEEE). http://dx.doi.org/10.1109/NOMS.2014.6838400
41. M. Torres Vega, S. Zou, D. Constantin Mocanu, E. Tangdiongga, A.M.J. Koonen, A. Liotta, End-to-end performance evaluation in high-speed wireless networks, in *Proceedings of the 10th International Conference on Network and Service Management*, Rio de Janeiro, Brazil, 17–21 Nov 2014 (IEEE)
42. D. Constantin Mocanu, G. Santandrea, W. Cerroni, F. Callegati, A. Liotta, Network performance assessment with quality of experience benchmarks, in *Proceedings of the 10th International Conference on Network and Service Management*, Rio de Janeiro, Brazil, 17–21 Nov 2014 (IEEE)
43. G. Exarchakos, V. Menkovski, A. Liotta, Can Skype be used beyond video calling? In *Proceedings of the 9th International Conference on Advances in Mobile Computing and Multimedia*, Ho Chi Minh City, Vietnam, 5–7 Dec 2011 (ACM)
44. M. Alhaisoni, A. Liotta, M. Ghanbari, Resource-awareness and trade-off optimization in P2P video streaming. Int. J. Adv. Media Commun. (special issue on High-Quality Multimedia Streaming in P2P Environments) **4**(1), 59–77 (Inderscience Publishers, 2010). doi:10.1504/IJAMC.2010.030005, ISSN:1741-8003
45. M. Alhaisoni, A. Liotta, Characterization of signalling and traffic in joost. J. P2P Netw. Appl. (special issue on Modelling and Applications of Computational P2P) **2** 75–83 (Springer, 2009). doi:10.1007/s12083-008-0015-5, ISSN:1936-6450
46. F. Agboma, A. Liotta, Addressing user expectations in mobile content delivery. J. Mob. Inf. Syst. (special issue on Improving Quality of Service in Mobile Information Systems), **3**(3), 153–164 (IOS Press, 2007)
47. D. Taubman, M. Marcellin, M. Rabbani, Jpeg 2000: image compression fundamentals, standards and practice. J. Electro. Imag. **11**(2), 286–287 (2002)
48. F. Pereira, T. Ebrahimi, *The MPEG-4 Book* (Prentice Hall, 2002)

49. D. Taubman, M. Marcellin, *JPEG2000: Image Compression Fundamentals, Practice and Standards* (Kluwer Academic Publishers, Massachusetts, 2002)
50. Z. Liu, L. Karam, A. Watson, Jpeg 2000 encoding with perceptual distortion control. IEEE Trans. Image Process. **15**(7), 1763–1778 (2006)
51. F.A. Kingdom, P. Whittle, Contrast discrimination at high contrasts reveals the influence of local light adaptation on contrast processing. Vision Res. **36**(6), 817–829 (1996)
52. J. Shapiro, Embedded image coding using zerotrees of wavelet coefficients. IEEE Trans. Signal Process. **41**(12), 3445–3462 (1993)
53. J. Li, Visual progressive coding, in *SPIE Proceedings Series* (Society of Photo-Optical Instrumentation Engineers, 1998), pp. 1143–1154
54. N. Kamaci, Y. Altunbasak, R. Mersereau, Frame bit allocation for the h. 264/avc video coder via cauchy-density-based rate and distortion models. IEEE Trans. Circuits Syst. Video Technol. **15**(8), 994–1006 (2005)
55. K. Ramchandran, A. Ortega, M. Vetterli, Bit allocation for dependent quantization with applications to multiresolution and mpeg video coders. IEEE Trans. Image Process. **3**(5), 533–545 (1994)
56. J. Ribas-Corbera, S. Lei, Rate control in dct video coding for low-delay communications. IEEE Trans. Circuits Syst. Video Technol. **9**(1), 172–185 (1999)
57. Y. Kim, Z. He, S. Mitra, A novel linear source model and a unified rate control algorithm for h. 263/mpeg-2/mpeg-4 in *Proceedings of IEEE International Conference on Acoustics, Speech, and Signal Processing, 2001. (ICASSP'01)*, vol. 3 (IEEE, 2001) pp. 1777–1780
58. Z. He, Y. Kim, S. Mitra, Low-delay rate control for dct video coding via ρ-domain source modeling. IEEE Trans. Circuits Syst. Video Technol. **11**(8), 928–940 (2001)
59. V. Menkovski, G. Exarchakos, A. Liotta, The value of relative quality in video delivery. J. Mob. Multimedia **7**(3), 151–162 (2011)
60. V. Menkovski, G. Exarchakos, A. Cuadra-Sanchez, A. Liotta, Estimations and remedies for quality of experience in multimedia streaming, in *Proceedings of the 3rd International Conference on Advances in Human-Oriented and Personalized Mechanisms, Technologies, and Services*, Nice, France, 22–27 Aug 2010 (IEEE)
61. V. Menkovski, A. Liotta, Intelligent control for adaptive video streaming, in *Proceedings of the International Conference on Consumer Electronics*, Las Vegas, US, 11–14 Jan 2013 (IEEE). http://dx.doi.org/10.1109/ICCE.2013.6486825
62. V. Menkovski, G. Exarchakos, A. Liotta, Tackling the sheer scale of subjective qoe, in *Mobile Multimedia Communications* (Springer, 2012), pp. 1–15
63. V. Menkovski, G. Exarchakos, A. Liotta, Machine learning approach for quality of experience aware networks, in *Proceedings of Computational Intelligence in Networks and Systems*, Thessaloniki, Greece, 24–26 Nov 2010 (IEEE)
64. V. Menkovski, A. Oredope, A. Liotta, A. Cuadra-Sanchez, Optimized online learning for QoE prediction, in *Proceedings of the 21st Benelux Conference on Artificial Intelligence*, Eindhoven, The Netherlands, 29–30 Oct 2009. http://wwwis.win.tue.nl/bnaic2009/proc.html, ISSN:1568–7805
65. A. Webster, C. Jones, M. Pinson, S. Voran, S. Wolf, An objective video quality assessment system based on human perception. SPIE Hum. Vis. Vis. Process. Digit. Display IV **1993**, 15–26 (1913)
66. N. Kanopoulos, N. Vasanthavada, R. Baker, Design of an image edge detection filter using the sobel operator. IEEE J. Solid-State Circuits **23**(2), 358–367 (1988)
67. J. Hu, H. Wildfeuer, Use of content complexity factors in video over ip quality monitoring, in *International Workshop on Quality of Multimedia Experience, QoMEx 2009* (IEEE, 2009) pp. 216–221
68. K. Seshadrinathan, R. Soundararajan, A. Bovik, L. Cormack, Study of subjective and objective quality assessment of video. IEEE Trans. Image Process. **19**(6), 1427–1441 (2010)
69. K. Seshadrinathan, R. Soundararajan, A. Bovik, L. Cormack, A subjective study to evaluate video quality assessment algorithms, in *SPIE Proceedings Human Vision and Electronic Imaging*, vol. 7527 (Citeseer, 2010)

70. T. Wiegand, G. Sullivan, G. Bjontegaard, A. Luthra, Overview of the h. 264/avc video coding standard. IEEE Trans. Circuits Syst. Video Technol. **13**(7), 560–576 (2003)
71. NTIA, Video quality metric (vqm) software (2012). http://www.its.bldrdoc.gov/resources/video-quality-research/software.aspx
72. K. Seshadrinathan, R. Soundararajan, A. Bovik, L. Cormack, *A Subjective Study to Evaluate Video Quality Assessment Algorithms*, vol. 7527 (2010)
73. V. Menkovski, A. Liotta, Adaptive psychometric scaling for video quality assessment. Signal Proces. Image Commun. (2012)

Chapter 3
Subjective QoE Models

Abstract The subjective QoE methods are concerned with quantifying the experienced quality of the users. Because these methods measure the subjective quality in an unmediated manner, their measurements are commonly used as 'ground truth' for evaluation of other methods [1]. In an end-to-end approach for QoE management, subjective estimation and subjective models are of key importance. It allows for evaluating the performance of the system and enables the loop back signal that closes the control loop. In this chapter these subjective QoE methods are discussed. QoE estimation methods based on rating, Just Noticeable Differences and difference scalling are presented as well as their computational counterparts that enable building of the QoE models.

3.1 Subjective QoE Approaches

In an end-to-end approach for QoE management [2–20] [21–29], subjective estimation and subjective models are of key importance. There are different approaches to subjective evaluation. The most direct technique is the rating approach, where the participants are asked to rate the quality of the content on different scales. The motivation here is that the participants can directly report the level to which their expectations have been met. However, research in psychophysics on measurements of subjective values demonstrates that the rating approach has significant drawbacks [30], mainly due to the high bias and variability in the results.

Other subjective evaluation methods such as the Just Noticeable Differences (JND) and the method of limits focus on estimating the parameter value for which the impairment becomes perceptible. The quality can be considered as acceptable as long as the change in parameters results in no perceivable difference using the method of limits [31]. In JND the smallest change in a parameter that results in detection is defined as 1 JND. Then the amplitude of the subjective value is measured using this unit, the JND. In this manner JND quantifies the amount the degradation of quality.

On the other hand, more recent research in psychophysical methods indicate that difference scaling methods show the best performance [32]. These methods

© Springer International Publishing Switzerland 2015   37
V. Menkovski, *Computational Inference and Control of Quality
in Multimedia Services*, Springer Theses, DOI 10.1007/978-3-319-24792-2_3

can deliver the relative differences in quality between different video samples and quantify the quality in a relative way [30].

The reason for the many different approaches in subjective quality measurements is mainly due to the difficulties associated with accurate measurement of subjective values [33]. In the following section we will examine the commonly used methods and present a more detailed discussion on the advancements we have made in this area.

3.2 State of the Art

In this section discussion on existing methods for video quality estimation is presented. The section is divided in two sub-sections, the first one focuses on methods that use rating, and the second focuses on the estimations with limits and noticeable differences.

3.2.1 Rating the Quality

The method most commonly encountered in the literature for subjective video quality evaluation is the rating method. This method has been standardized by the International Telecommunication Union (ITU) in [34]. In the recommended setup for a subjective study the viewing conditions are strictly controlled, 15 or more non-expert participants are selected and trained for the exercise. The grading scale for absolute category rating (ACR) is defined as Mean Opinion Score (MOS) and it has five values: 1—'Bad', 2—'Poor', 3—'Fair', 4—'Good', 5—'Excellent'. Similarly for impairments: 'Very annoying', 'Annoying', 'Slightly annoying', 'Perceptible' and 'Imperceptible'. There is also a comparisons scale for Differential MOS (DMOS) going from 'Much worse' to 'Much better' with an additional value in the middle 'The same'.

A typical example of a subjective assessment using ACR has been executed in [35]. The assessment is on quality degradation of error-prone network transmission. The subjective study was executed on 40 subjects in order to provide for a database for further evaluation of FR, RR and NR objective VQA.

The effect of interactions between bit-rate levels and temporal freezes in video playback on the quality have been evaluated in [36] also by means of an ACR subjective study. They concluded that the video quality is affected by both types of impairments, however the temporal impairment have a more intensive effect. Furthermore, introducing the second impairment affects the effect on the perceived quality from the first. They developed a non-linear model integrating both impairments to predict the overall quality.

The Live video database [37] contains a set of test sequences impaired with two types of compression methods and two types of transports errors. In this study the subjective test was implemented using the DMOS [38] scale. The study was executed to evaluate the performance of many objective VQA methods, using subjective data as benchmark. Thus, the objective methods (PSRN, SSIM, MS-SSIM, …, MOVIE) were evaluated by their correlation with the subjective data. The results indicate that the MOVIE VQA index delivers the best performance overall.

The Video Quality Experts Group (VQEG) reports the results of their recent subjective study of High Definition Television (HDTV) in [39]. They explored the effects that compression quantization level, bit-rate, transmission errors and different concealment strategies have on the perceived quality. They produced a linear model of the video quality and the impairment factors using two or ten parameters.

Evidently the rating method is widely used. Even though, the method gives a sense of direct measurement of the human perceived quality, it is based on a flawed concept. This type of 'direct' psychophysical measurements dates back to the work of Stevens (1957) [40]. Stevens claims that there are internal psychological scales that can be empirically measured by directly inquiring the 'how much' question. However, later work in psychophysics uncovered that such direct methods are inheritably biased due to the qualitative nature of the scale ('Poor', 'Bad', 'Fair') and present a high level of variance [41–44]. In fact, Shepard continues states that: "Although the (human) subject himself can tell us that one such inner magnitude is greater than another, the psychophysical operations that we have considered are powerless to tell us anything further about 'how much' greater one is than the other." [45].

In much later work, similar conclusions have been reached by video quality experts. Watson concurs that the brain perceptual system is more accurate at grasping 'differences' rather than giving absolute rating values [46]. In his analysis of the properties of subjective rating, Winkler discusses the MOS variability and standard deviation in a set of subjective databases [47]. The analysis shows that the standard deviation of MOS for the midrange is between 15 and 20 % of the scale, although it decreases at the edges. This is confirmed also by a recent study by the video quality experts group [48, 49]. The diagrams depict the standard deviation for Mean Opinion Score (MOS) rating tests and Differential Mean Opinion Score (DMOS) tests (Fig. 3.1). The variance is presented on the vertical axis as a percentage of the rating scale.

It is evident that the variance of the results is in the range of 15–20 % of the scale. Particularly the middle range of the scale has variance reaching 20 % and above. Thus, if we aimed to map the MOS scale to quality labels such as 'Poor' (lower end of the x scale), 'Fair' (on the middle) and 'Good' (high end of the scale); the results would hardly be reliable. These diagrams give a general feel as to the actual perceived quality but cannot be used to directly map quality labels onto a continuous axis or to draw any conclusive results [50]. Furthermore, the question remains as to whether the distance between the 'Poor' and 'Fair' value on the x-axis is the same as the distance between the 'Fair' and 'Good' perception of the participant.

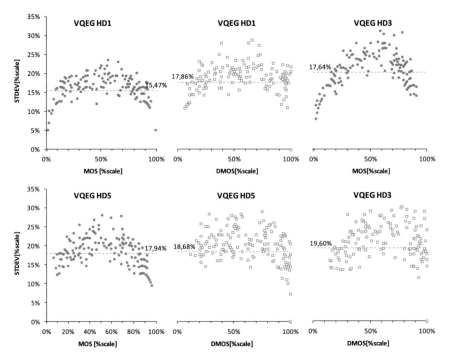

Fig. 3.1 Subjective variability [48]

An analysis of the subjective scales in [51] presents more interesting perspectives on the rating method. A striking one is that because the MOS scale is an ordinal qualitative scale it should not be used as a quantitative scale. The authors present an argument that the commonly used mapping in the literature from qualitative to a quantitative scale in rating is not justified. Therefore analysis of MOS in decimal values would be invalid as well as its variability calculated to less than one single ordinal steps in the scale. In other words, numbers associated with the 5 labels do not represent actual distance between them. To avoid these issues with the scales, the authors propose a label-free scale or labels only at the end of the scale. This approach, however convenient for the analysis of the results, opens the question as to how people would map their internal representation of 'Good' and 'Bad' on this label-free scale. After all the main goal of the rating is to deliver a qualitatively measurable result such as 'Good' or 'Bad'.

3.2.2 Limits and Noticeable Differences

Due to the limitations of rating, other approaches have been developed for measuring the quality or degradation in video services.

The method of limits is a psychometric approach proposed by Fechner [52]. The method is designed to determine threshold level of stimuli by introducing a gradual increase until the stimulus becomes detectable. The procedure can also be run in reverse (until the stimulus becomes undetectable). The participant gives a 'Yes' or 'No' response at each level of intensity according to whether the change has been detected.

The method has been also used for determining the acceptability of video quality. In [53], MecCarthy et al. use the method of limits to explore the effects of encoding quantization and changes in frame-rate on the acceptability of video quality. They implemented the study on two devices, a desktop computer and a hand held palmtop computer. The study collected data consisting of acceptability ratings for the different test conditions. These acceptability ratings were transformed into ratio measures by calculating the proportion of time during each 30-s period that the quality was rated as acceptable. The study was implemented on sports coverage videos with 41 participants on the first type of device and 37 on the second. The results of this study indicate that the participants are more sensitive to reduction in frame quality (quantization) than to changes in frame-rate. The authors claim that the results challenge the conventional wisdom that for sport events with high amount of movement, high frame-rate is necessary for high level of quality.

In a subjective study addressing the user expectations in mobile content delivery [31] the authors examined the acceptability of QoE, for different types of content and on three different devices, using the method of limits. Content types included: news broadcast, sports, video game animations, music videos and movies. They repeated the experiment in an ascending and descending order. In this study they evaluated the quality based on the video encoding bit-rate and the frame-rate of the video. The results collected from 96 participants varied significantly over different types of content. On the mobile phone terminal, the mean acceptability thresholds for football were found to be 128 kb/s with 15 frames/s, while for the Romance movie is 32 kb/s with 10 frames/s. Even though, this finding about the big difference in quality versus resources for different types of content is most interesting, the results bring light to one of the main pitfalls of this method. The method of limits presents a significant histeresis in the results, the finding on the ascending order vary significantly from the finding on the descending order evaluation. Because of this effect, accurate estimation of the acceptability is not clear for the values within the hysteresis. This is the main drawback of this method.

The JND as introduced by Weber [54] is defined as the smallest detectable difference between two intensities of a sensory stimulus. It is a statistical value usually defined as the level detectable in 50 % of the tested cases. The concept of JND was re-introduced for scaling video quality by Watson in a proposal for a new quality scaling method [46]. In [55], Watson and Kreslake execute a subjective study by asking the participants which of two presented videos is more impaired. This is called 'pair comparison'. From the responses to that simple question, they measure the observer's internal 'perceptual scale' for visual impairments. This method relies on

Thurstone's 'Law of comparative judgment' [56]. Thurstone proposes that physical stimuli give rise to perceived magnitudes on a one-dimensional internal psycholog-ical scale. He continues to include an inevitable variability in the neural system. In 'Case five' of the law, the stimuli are perceived with a normal distribution with a standard deviation of 1. So if two stimuli are presented and the participant is asked to discriminate between the two, the probability of giving the right question is a function of the distance between the mean values of both probability distributions on the internal scale. In this manner by acquiring enough data the most likely values for different stimuli can be inferred statistically.

However, when participants are presented with a two different levels of quality in video, they will discriminate between two points on the internal scale. If the two points are not close enough, discriminating becomes too easy and leads to results that tend to sort this points on the intensity scale, but not scale them. In other words results from pair-wise comparison do not yield information about the distance between the points, only their order on the scale. Relying on the probability of incorrectly responding to the levels of quality is challenging and requires the ability to very finely tune the parameters of the video. This is not trivial, as encoding parameters usually change in discrete steps. Furthermore, subjective methods are commonly used to estimate the quality of a predetermined set allowed (or favourable) parameter values. For example 10 different levels between 64 kb/s and 2 Mb/s of constant bit-rate encoding, cannot be directly scaled by the proposed method without some kind of interpolation between the samples. However, the effects of the interpolation on the perceived quality in light of the complex masking effects of the HVS are not analyzed.

On the other hand if the participants are presented with two ranges of quality (such as in two pairs of videos) and asked to discriminate between the two ranges, then the intensity of quality is actually scaled. This approach is discussed in more detail in the following section.

3.3 Maximum Likelihood Difference Scaling

The Maximum Likelihood Difference Scaling method (MLDS) is a psychophysical method that scales the relative differences perceived psychologically between phys-ical stimuli [57]. In other words by executing a 2-alternative forced choice (2AFC) subjective study and statistical analysis, the method will output a relative scale of the difference between stimuli with increasing or decreasing intensities. The MLDS method has been used by Maloney and Yang as a tool for subjective analysis of image quality [57]. Furthermore, Charrier et al. show how to use MLDS to achieve differ-ence scaling of compressed images with lossy image compression techniques using MLDS [58]. They implement a comparison of image compression in two different colour spaces, and conclude that in the CIE 1976 L*a*b* colour space the images can be compressed by 32 % more, without additional loss in perceived quality. Their

Fig. 3.2 Psychometric function

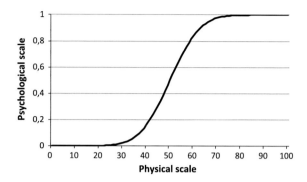

results and discussion clearly show the applicability of MLDS and the ease of collecting data with it.

The image quality study clearly presents the advantages of using difference scaling methods for applications where quality is the target measurement. Motivated by the advantages of this approach, we have developed the appropriate tools for implementation of difference scaling for estimation of quality in video.

We carry out a subjective study to estimate the quality scale for a range of videos with different spatial and temporal characteristics. The results presented demonstrate that MLDS can be used for estimating quality of video with higher accuracy and significantly lower testing costs than subjective rating.

As discussed above, the goal of the MLDS method is to map the objectively measurable scale of video quality to the internal psychological scale of the viewers. The output is a quantitative model for this relationship based on a psychometric function [32] as depicted in Fig. 3.2.

The horizontal axis of the Fig. 3.2 represents the physical intensity of the stimuli—in our study this will be the video bit-rate. The vertical axis represents the psychological scale of perceived difference in signal strength—for our purpose the difference in video quality. The perceptual intensity of the first (or reference) sample ψ_1 is 0 and the last sample perceptual difference ψ_10 is fixed to 1 without the loss in generality [59]. The MLDS produced model is an estimate of the rest of the amplitudes of the stimuli on viewers' internal quality scale.

The 2AFC test is designed in the following manner. Two pairs of videos are presented to the viewers (ψ_i, ψ_j and ψ_k, ψ_l). The intensity of the physical stimuli is always in the following manner $i < j$ and $k < l$. The method needs to compare sizes of distances between the qualities of videos so that the results can directly let us build a model of the quality distance between all of the presented videos.

The viewer needs to select the pair of videos that have bigger difference in quality between them. In other words if the expression $|\psi_j - \psi_i| - |\psi_l - \psi_k| > 0$ is true the viewer selects the first pair, otherwise he or she will choose the second.

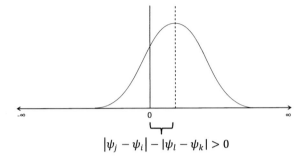

Fig. 3.3 Signal of 1 unit superimposed over noise with 0 mean and standard deviation of 1

$$|\psi_j - \psi_i| - |\psi_l - \psi_k| > 0$$

Because the stimuli are ordered as $i < j$ and $k < l$ we can safely assume due to the monotonicity of the psychometric curve that in the psychological domain also $\psi_j \geq \psi_i$ and $\psi_l \geq \psi_k$ and we drop the absolute values. The decision variable used by the observer is the following:

$$\Delta(i, j, k, l) = \psi_j - \psi_i - \psi_l + \psi_k + \epsilon \tag{3.1}$$

where ϵ is the error or noise produced by the viewers visual and cognitive processing. As defined in (3.1) the observer will select the first pair when $\Delta(i, j, k, l) > 0$ or the second one when $\Delta(i, j, k, l) < 0$.

In order to use the maximum likelihood method to determine the $\Psi = (\psi_1, \ldots, \psi_1 0)$ parameters, we need to define the likelihood (probability given the parameters) that the viewer will find the first pair with larger difference than the second pair. For this the method models the perceived distances using signal detection theory (SDT) [60].

The equal variance Gaussian model defined in the SDT is used to model the process of selection that the user is executing for each presented pair. This model assumes that the signal is contaminated with ϵ, a Gaussian noise with zero mean and standard deviation of σ (Fig. 3.3). Each time the observer is presented with a pair of videos, the perceived difference is a value of the random variable X drawn from the distribution given in Fig. 3.3. The distribution in Fig. 3.3 is with arbitrary signal strength of 1.

The probability that $\Delta(i, j, k, l) > 0$ is given by the surface under the Gaussian from zero to plus infinity (Fig. 3.4). For reasons of mathematical simplicity it is better to represent the surface under the curve with a cumulative Gaussian function. The inverse portion of the surface (Fig. 3.5) is as in Eq. 3.2.

$$F(x; \mu, \sigma^2) = \Phi\left(\frac{x - \mu}{\sigma}\right) = \frac{1}{\sqrt{2\pi}} \int_x^{-\infty} e^{\frac{-(t-\mu)^2}{2}} dt \tag{3.2}$$

Fig. 3.4 The shaded area corresponds to the probability that the signal is positive

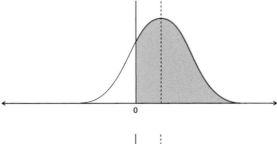

Fig. 3.5 The shaded area corresponds to the probability that the signal is negative

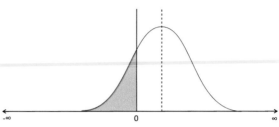

Looking at the inverse part of the surface under the Gaussian the probability of detecting the signal would be:

$$P(R = 1; \mu_s, \sigma^2) = 1 - \Phi\left(\frac{0 - \mu_s}{\sigma}\right) = \Phi\left(\frac{\mu_s}{\sigma}\right) \tag{3.3}$$

and

$$P(R = 0; \mu_s, \sigma^2) = 1 - P(R = 1; \mu_s, \sigma^2) = 1 - \Phi\left(\frac{\mu_s}{\sigma}\right) \tag{3.4}$$

where μ_s is the mean or the intensity of the signal, σ is the standard deviation of the noise and R is 1 when the first pair is selected and 0 when the second pair is selected. The likelihood for the whole set of responses in a test is the product of all of the individual probabilities where $\delta(i, j, k, l)_n = \psi_{j_n} - \psi_{i_n} - \psi_{l_n} + \psi_{k_n}$.

The Maximum Likelihood method estimates the parameters, such that the given likelihood is maximized. For example, if we have $x = \{x^t\}(t = 1 \ldots N)$ instances drawn from some probability density family $p(x|\theta)$ defined up to parameters θ (Eq. 3.5).

$$x^t \sim p(x|\theta) \tag{3.5}$$

If the x^t samples are independent, the likelihood parameter θ given a sample set x is the product of the likelihood of individual points (Eq. 3.6).

$$L(\theta|x) \equiv p(x|\theta) = \prod_{N}^{t=1} p(x^t|\theta) \tag{3.6}$$

Fig. 3.6 Video display layout in MLDS application

There is no closed form for such a solution, so a direct numerical optimization method needs to be used to compute the estimates (Eq. 3.7).

$$\widehat{\theta} = argmax_{\theta} l(\theta|x) \tag{3.7}$$

3.3.1 The Video Subjective Study

The experimental setup consists of a web application that displays two pairs of videos to the viewer as shown in Fig. 3.6. The user response is recorded in the application database. The web application is developed using the java server pages technology [61]. The videos are displayed using the JW player [62], which is a Flash 5 web player capable of displaying H.264 encoded videos. The videos are encoded using the X264 library [63] and saved in mp4 file format.

The raw videos are the unimpaired samples of the Live video database used in the objective VQA.

Ten different videos are encoded at constant bit-rate, each one at different values ranging from 2 Mbps to 64 kbps. The frame-rate is 25 and the spatial resolution is 768 by 432 pixels. The video player is configured to pre-buffer the full content before playing, so additional impairments such as freezes during the playback are avoided.

The results are collected in a database in the format:

bit-rate 1	bit-rate 2	bit-rate 3	bit-rate 4	R (index bigger pair)

For computing the difference scales ($\Psi = (\psi_1, \ldots, \psi_{10})$), we used the MLDS implementation [59] in R programming language [64]. The output ψ values are fitted to a psychometric curve using a probit regression fit with variable upper/lower asymptotes using the 'psyphy' package in R [65].

The results of the subjective study are the parameters μ and σ of a cumulative Gaussian (psychometric) curve that describes dependency between the QoE and bit-rate. This curve is calculated for each type of video assesed during the study.

3.3.2 MLDS Subjective Results

The MLDS experiment with 10 levels of stimuli requires 210 responses to cover all possible combinations for a single video. We have done three rounds per video sample or 630 tests for each video; in total we have collected 6300 responses. The videos are displayed one at the time or in pairs. They are 10 s long, so to view a single test up to 40 s are needed, but in most cases the larger difference is evident much sooner to most observers.

To calculate the standard error we executed a bootstrap [66] fitting procedure with 10,000 rounds. The mean values are given in Fig. 3.7 and the standard deviation for each point in Fig. 3.8.

The results in Fig. 3.7 show that most of the videos follow a similar trajectory of the difference in quality. There is little perceived difference down to 512 kbps and then a rapid rise appears. The difference is not zero in the high range, as we can also see from the standard error on the points from 1536 to 512 kbps, but it is very low relative to the lower bit-rate samples. This means it is safe to say that there is little benefit from increasing the bit-rate above 512 kbps. The exception is the 'rb' video

Fig. 3.7 Results of the MLDS experiment by video type

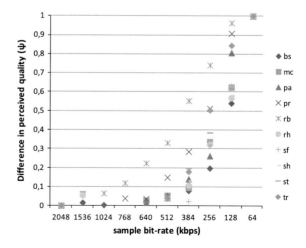

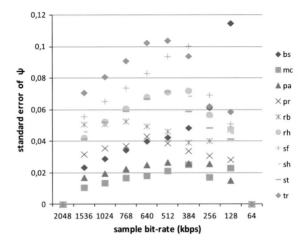

Fig. 3.8 Standard error of the MLDS results by video type

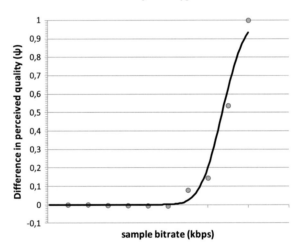

Fig. 3.9 Fitting a cumulative Gaussian on the 'bs' MLDS model

and somewhat the 'pr' video. The 'rb' video displays a surface of water, which shows significantly different compression characteristics than the rest of the videos.

To quantitatively analyze the characteristics of each model we fitted a cumulative Gaussian curve to each difference model as demonstrated in Fig. 3.9, which represents the psychometric curve [67]. There is high goodness of fit to the curve with small residuals. This further demonstrates the success of this subjective study to model the quality difference perception with a psychometric curve.

For each video the μ and σ of the fitted curve are given in Fig. 3.1. A plot of each of the fitted models is given in Fig. 3.10. The plotted curves model a smooth quality distance for different bit-rates from the reference 2Mbps video.

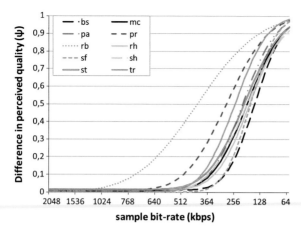

Fig. 3.10 Fitted psychometric curves from the MLDS results

Table 3.1 The μ and σ of the cumulative Gaussian

	bs	mc	pa	pr	rb	rh	sf	sh	st	tr
μ	−5.43	−5.07	−5.08	−4.57	−4.09	−5.22	−5.54	−5.13	−5.00	−4.94
σ	0.24	0.20	0.20	0.15	0.11	0.22	0.25	0.21	0.20	0.19

Observing the parameter values in Table 3.1 we can draw the same conclusions as above, in a quantitative form. The mean value of the psychometric curve of the 'rb' video is noticeably lower than the rest of the videos, so its curve gradient increases earlier than the other video types. The remaining psychometric curves cluster together and confirm that most of these videos difference in quality is negligible to the reference, down to 512 kbps, while the bit-rate between 256 and 128 kbps is half way to the perceived distance between the reference and the 64 kbps video. The results accurately capture the nonlinearity in the perceived quality by the viewers.

3.4 Adaptive MLDS

The MLDS method is appealing for its simplicity and efficiency, but is intrinsically not scalable as it considers all combinations of the samples. For instance, one full round of tests for ten levels of stimuli (i.e. video qualities) requires 210 individual tests. We developed an optimized version of MLDS, referred to as adaptive MLDS [48, 68] to reduce the redundancy of conventional MLDS tests whilst also maintaining the reliability of the results. The strategy of adaptive MLDS is to employ active learning to minimize the number redundant tests.

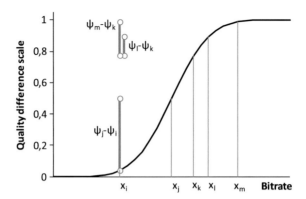

Fig. 3.11 Monotonicity of the psychometric curve

3.4.1 Adaptive MLDS Method

This method is based on the idea that with the knowledge acquired by executing a small number of tests we can already estimate the answers of the remaining tests. Then using these estimates together with the known responses we execute the MLDS method. Executing the MLDS with more responses helps the argument maximization procedure to produce more stable solutions. The estimation of the unanswered tests is based on the characteristics of the psychometric curve.

The idea comes from the notion that some of the tests are covering the range of others. In fact, all of the tests overlap with others in one way or another. The approach makes use of the intrinsic characteristics of the psychometric curve, a monotonically increasing function $\vec{\Psi} = f(\vec{X})$. Consequently, for $k < l < m$, $x_k > x_l > x_m$, if $x_k - x_l > x_k - x_m$ in the physical domain then $\psi_k - \psi_l \geq \psi_k - \psi_m$ in the psychological domain (Fig. 3.11).

If we now observe five samples x_i, x_j, x_k, x_l, x_m such that $i < j < k < l < m$ and we observe two tests $T_1(x_i, x_j; x_k, x_l)$ and $T_2(x_i, x_j; x_k, x_m)$, the perceived qualities in the psychological domain are $\psi_i < \psi_j < \psi_k < \psi_l < \psi_m$. If in T_2 the first pair is bigger or $\psi_j - \psi_i > \psi_m - \psi_k$ that would mean that $\psi_j - \psi_i > \psi_m - \psi_k \geq \psi_l - \psi_k$. In other words, if in T_2 the first pair is selected with a bigger difference, then in T_1 the first pair has a bigger difference as well (Fig. 3.11). There are many different combinations of tests that have this dependency for the first pair or the second pair. We can generate a list of dependencies for each pair, based on two simple rules:

- Let us assume test $T_1(a, b, c, d)$ such that $a < b < c < d$, $\psi_b - \psi_a > \psi_d - \psi_c$ and test $T_2(e, f, g, h)$ with $e < f < g < h$. If $e \leq a < b \leq f$ and $c \leq g < h \leq d$ then $\psi_f - \psi_e > \psi_h - \psi_g$ (Fig. 3.12).
- Let us assume test $T_1(a, b, c, d)$ such that $a < b < c < d$, $\psi_b - \psi_a < \psi_d - \psi_c$. If for test $T_2(e, f, g, h)$ with $e < f < g < h$ the following holds: $a \leq e < f \leq b$ and $g \leq c < d \leq h$ then $\psi_f - \psi_e < \psi_h - \psi_g$ (Fig. 3.13).

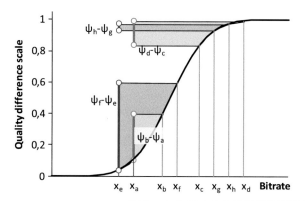

Fig. 3.12 If first pair in T1 is bigger then first pair of T2 is bigger as well

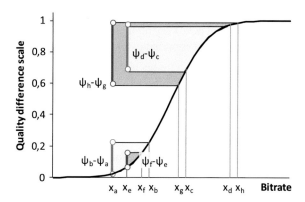

Fig. 3.13 If second pair in T1 is bigger then second pair of T2 is bigger as well

Thus, after introducing an initial set of responses we can start estimating the probabilities of the rest. However first we need to learn the probabilities of each of the known responses to be actually valid. MLDS estimates the values of the psychological parameters $\Psi = (\psi_1, \ldots, \psi_{10})$ such that the combined probabilities of each response or the overall likelihood of the dataset is maximized. Nevertheless, after the argument maximization is finished the different responses have different probabilities of being true. Having a set of initial Ψ quality values as the prior knowledge about the underlying process coming from the data, we generate the estimations for the rest of the tests. The interdependencies from the tests are far more complex, of course. Let us assume, for example, a test T_1 that depends on tests T_2 and T_3. If the answer from T_2 indicates that the first pair has a larger difference in T_1 and the answer from T_3 indicates the opposite, then we need to calculate the combined probability of T_2 and T_3 to estimate the answer of T_1. Assuming that the responses of T_2 and T_3 are independent and that the probability of giving the first and second answer is the same, the combined probability of T_2 and T_3 is given by Eq. 3.8.

$$P(T_1) = \frac{P(T_2)(1 - P(T_3))}{P(T_2)(1 - P(T_3)) + (1 - P(T_2))P(T_3 f)} \tag{3.8}$$

Of the remaining tests that have no responses, some will have higher estimates than others. In other words, we have better estimations for some of tests than others. To improve the speed of learning, the adaptive MLDS process focuses on tests that have smaller confidence in the estimations. This way when we receive the next batch of responses, the overall uncertainty in the estimates should be minimized. The goal of the adaptive MLDS is to develop a metric that will indicate when the amount of tests is sufficient for determining the psychometric curve. We can obtain this indication from the probabilities of the estimations. As we get more responses by asking the right questions, the estimation for the rest of the tests keeps improving. At some point adaptive MLDS will have very high probabilities of estimating correctly all of the remaining tests. This is a good indication that no more tests are necessary.

3.4.2 Learning Convergence

The test estimation of adaptive MLDS provides a good indication for concluding the experiment. When the confidence of the estimations of the remaining tests becomes high, the probability of a surprise in the future responses of the participants goes down. With this, the need for further tests also becomes smaller. Even so, an indication of the amount of surprise from the participant responses would be useful to determine the utility of more testing.

Each batch of new data that is collected has a specific amount of information gain at each point in the experiment. The amount of information gain is proportional to the amount of surprise that the data delivered, i.e. how much the data changed the existing model. In other words when we receive responses that are completely expected our model of the differences in quality will not change at all. These responses bring no surprise and their information gain is zero. However, if the responses change our belief about the scaled quality and the model changes, then the new data has resulted in information gain. The information gain calculation is based on the Kullback-Leibler (KL) divergence [69]. The KL divergence is a way of comparing two probability distributions and produces the number of average bits that need to be used to explain this difference [70]. Using the KL divergence, the information gain in bits coming from data that change the model's distribution with mean and standard deviation to a model distribution with and is given in Eq. 3.9.

$$I = \log_2 \frac{\sigma_0}{\sigma_1} + k\left(\left(\frac{\mu_1 - \mu_0}{\sigma_0}\right)^2 + \frac{\sigma_1^2 - \sigma_0^2}{\sigma_0^2}\right) \tag{3.9}$$

The information gain is giving us a tool to determine when the learning process has converged, i.e. when additional tests would not bring any better understanding of the problem.

3.4.3 Experimental Setup and Results

To demonstrate the performance of 'adaptive MLDS', we have developed a software test-bed. The software simulates the learning process of the adaptive MLDS algorithm by sequentially introducing data from our earlier subjective study [30]. The simulation test-bed is a Java application that loads the subjective data from a file, and then sequentially introduces new data-points. The data-points are selected by the adaptive MLDS algorithm and the estimated values are used to calculate the video quality scaling in each iteration. The output is compared to the output of running MLDS on the full dataset. The root mean square error (RMSE) is computed on the differences. In parallel, a random introduction of data is also executed as a baseline for comparison. The adaptive MLDS algorithm is implemented in Java, while the MLDS software from [59] is used directly from R using a Java to R interface. To account for the variation in the results due to the random start and random data introduction in the comparison process, the simulation is repeated 100 times and results are averaged. The simulation process was computationally very demanding. Each numerical optimization was bootstrapped 1000 times. This was repeated for each step in the introduction of new batch of data and for each video. All this for a single simulation. To handle the computational demand the simulation was executed on a high-performance computing grid. Adaptive MLDS as an active learning algorithm explores the space of all possible 2AFC tests with the goal of optimizing the learning process. It also provides an indication of confidence in the model built on the subset of the data, which provides for early stopping of the experiment. The performance of adaptive MLDS is presented in Figs. 3.14, 3.15, 3.16, 3.17 and 3.18. Figure 3.14 shows the accuracy of the estimations for ten types of videos against the number of introduced data-points. This should be compared with plain MLDS whereby all 210 tests must be performed. As it is evident for all types of videos the accuracy of the estimations is very high. With most videos we have $>95\%$ accuracy with just 15 tests. The Station and River bed videos show lower accuracy than the rest, but they recover quickly above 90% when around 90–100 data-points are presented. This indicates that for most videos we can estimate all of the tests accurately after just about 60 responses, although some videos require about 100 tests. In Fig. 3.15 we observe the accuracy of the model generated with adaptive MLDS compared to the model of the classical MLDS. The horizontal axis represents the number of points introduced at the time the calculation was executed; the vertical axis gives the RMSE (root mean square error) between the estimated values and the values computed on the whole dataset. We can clearly observe that the adaptive MLDS model differs from the 'true' model (learned from the full data set) much less than the model built by introducing data randomly. In Fig. 3.16 we present the standard deviation of the

Fig. 3.14 Accuracy of the estimations

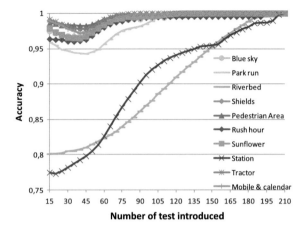

different value for the RMSE at each point. The results in this figure further support the fact that adaptive MLDS produces superior results in terms of error and variability. Figure 3.17 presents the distribution of the confidence or the probabilities of those estimations. The vertical axis the number of unanswered tests is plotted with different darkness at each step of the experiment. The ratio of the shadings represents the distribution of the confidence in the estimation of those tests. Starting from the initial 15 data points most of the unknown 195 tests are estimated with 0.5 accuracy. But soon after introducing more data, the estimations rapidly improve. Between 40 and 60 collected answers the confidence in the estimations was close to 1, suggesting that the rest of the tests are not necessary and that we can correctly estimate the psychometric curve without them, which reinforces our claims as per Fig. 3.15. The accuracy of the predicted psychometric curves is high for all datasets in this range. The RMSE is below 0.3 predicted values. The accuracy in the prediction is generally very high and improves with the introduction of more data, as shown in Fig. 3.14. As expected, the Riverbed and Station videos are more difficult to learn due to high noise in the answers, which makes them also more difficult to estimate. Finally we look at the information gain for each of the videos on each step of the experiment in Fig. 3.18. The results of the information gain analysis confirm that perturbations in the model were occurring only in the beginning of the experiment. Most of the videos had very small information gain after the 60th response had been received. The only exception is the station video, where the model changes even after most of the responses have been received. The results for all the other videos concur with the RMSE values of Fig. 3.15. However, the information gain criterion also considers the changes in the standard deviation of the model, which are not captured by the RMSE. This decrease in the variation of the model of the Station video explains why there is still information gain even though the mean RMSE is low.

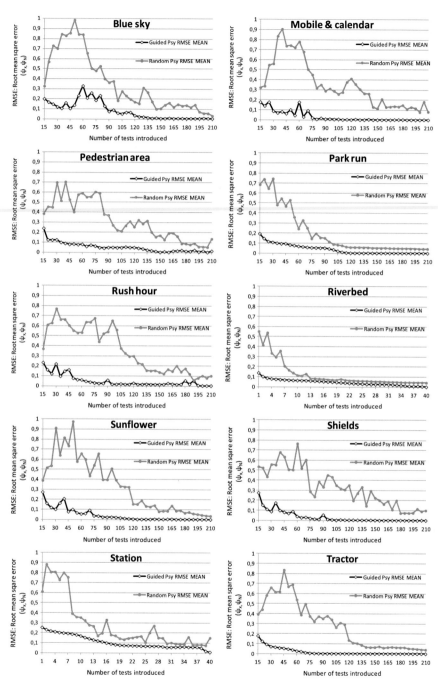

Fig. 3.15 Mean RMSE for the ten types of video

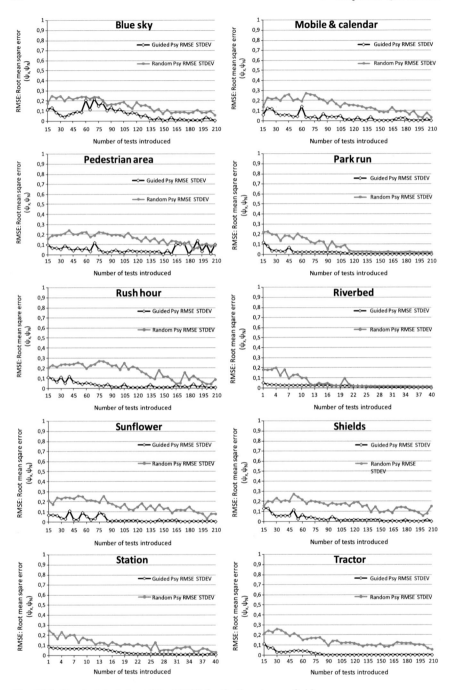

Fig. 3.16 Standard deviation of the RMSE for the three types of video

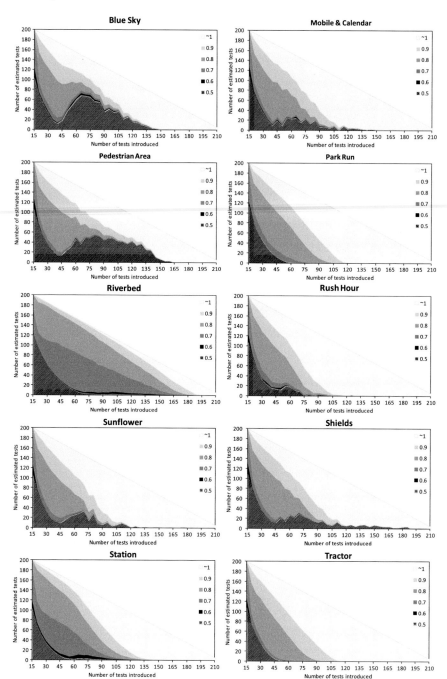

Fig. 3.17 Estimation confidences for the three types of videos over the number of introduced data-points

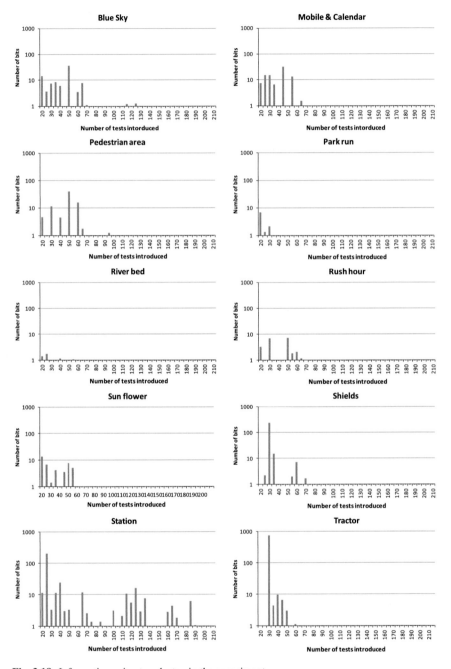

Fig. 3.18 Information gain at each step in the experiment

3.5 Conclusions

Video quality is one of the key factors of the overall QoE of video enabled network services [71, 72]. Moreover, due to its intensive footprint on the resources [73], video quality is also a key aspect for efficient QoE aware management of these services [74]. Subjective VQA is a slow and expensive process, but it is a necessary component of QoE management because it forms the basis for validating any objective quality assessment procedures. In this chapter we have presented the state of the art of subjective VQA and our contribution to the area. Rating is still the prevalent methodology for subjective VQA, however in light of the advantages of difference scaling methods more development in this area is expected. In the following chapters the important role of subjective QoE models is further demonstrated, as they take a key role in management of multimedia services.

References

1. S. Winkler, P. Mohandas, The evolution of video quality measurement: from PSNR to hybrid metrics. IEEE Trans. Broadcast. **54**(3), 660668 (2008). http://ieeexplore.ieee.org/xpls/abs_all.jsp?arnumber=4550731
2. G. Exarchakos, V. Menkovski, L. Druda, A. Liotta, Network analysis on Skype end-to-end video quality. Int. J. Pervasive Comput. Commun. **11**(1) (2015). http://www.emeraldinsight.com/doi/abs/10.1108/IJPCC-08-2014-0044
3. G. Exarchakos, L. Druda, V. Menkovski, P. Bellavista, A. Liotta, Skype resilience to high motion videos. Int. J. Wavelets Multiresolut. Inf. Process. **11**(3) (2013). http://dx.doi.org/10.1142/S021969131350029X
4. F. Agboma, A. Liotta, Quality of experience management in mobile content delivery systems. J. Telecommun. Syst. special issue on the Quality of Experience issues in Multimedia Provision, **49**(1), 85-98 (2012). doi:10.1007/s11235-010-9355-6
5. V. Menkovski, G. Exarchakos, A. Liotta, A. Cuadra Sánchez, Quality of experience models for multimedia streaming. Int. J. Mobile Comput. Multimedia Commun. **2**(4), 1–20 (2010). http://www.igi-global.com/ijmcmc/. doi:10.4018/jmcmc.2010100101. ISSN: 1937-9412
6. M. Alhaisoni, A. Liotta, Characterization of signalling and traffic in Joost. J. P2P Networking Appl. special issue on Modelling and Applications of Computational P2P, **2**, 75–83 (2009). doi:10.1007/s12083-008-0015-5, ISSN: 1936-6450
7. M. Torres Vega, E. Giordano, D.C. Mocanu, D. Tjondronegoro, A. Liotta, Cognitive no-reference video quality assessment for mobile streaming services, in *Proceedings of the 7th International Workshop on Quality of Multimedia Experience*, Messinia, Greece, 26–29 May 2015 (IEEE). http://www.qomex.org
8. A. Liotta, L. Druda, G. Exarchakos, V. Menkovski, Quality of experience management for video streams: the case of Skype, in *Proceedings of the 10th International Conference on Advances in Mobile Computing and Multimedia*, Bali, Indonesia, 3–5 Dec 2012 (ACM). http://dx.doi.org/10.1145/2428955.2428977
9. V. Menkovski, A. Liotta, Intelligent control for adaptive video streaming, in *Proceedings of the International Conference on Consumer Electronics*, Las Vegas, US, 11–14 Jan 2013 (IEEE). http://dx.doi.org/10.1109/ICCE.2013.6486825

10. A. Liotta, D. Constantin Mocanu, V. Menkovski, L. Cagnetta, G. Exarchakos, Instantaneous video quality assessment for lightweight devices, in *Proceedings of the 11th International Conference on Advances in Mobile Computing and Multimedia*, Vienna, Austria, 2–4 Dec 2013 (ACM). http://dx.doi.org/10.1145/2536853.2536903

11. D.C. Mocanu, A. Liotta, A. Ricci, M. Torres Vega, G. Exarchakos, When does lower bitrate give higher quality in modern video services?, in *Proceedings of the 2nd IEEE/IFIP International Workshop on Quality of Experience Centric Management*, Krakow, Poland, 9 May 2014 (IEEE). http://dx.doi.org/10.1109/NOMS.2014.6838400

12. D.C. Mocanu, G. Exarchakos, A. Liotta, Deep learning for objective quality assessment of 3D images, in *Proceedings of IEEE International Conference on Image Processing*, Paris, France, 27–30 Oct 2014 (IEEE)

13. M. Torres Vega, S. Zou, D. Constantin Mocanu, E. Tangdiongga, A.M.J. Koonen, A. Liotta, End-to-end performance evaluation in high-speed wireless networks, in *Proceedings of the 10th International Conference on Network and Service Management*, Rio de Janeiro, Brazil, 17–21 Nov 2014 (IEEE)

14. D. Constantin Mocanu, G. Santandrea, W. Cerroni, F. Callegati, A. Liotta, Network performance assessment with quality of experience benchmarks, in *Proceedings of the 10th International Conference on Network and Service Management*, Rio de Janeiro, Brazil, 17–21 Nov 2014 (IEEE)

15. D. Constantin Mocanu, G. Exarchakos, H.B. Ammar, A. Liotta, Reduced reference image quality assessment via Boltzmann machines, in *Proceedings of the 3rd IEEE/IFIP IM 2015 International Workshop on Quality of Experience Centric Management*, Ottawa, Canada, 11–15 May 2015 (IEEE)

16. M. Torres Vega, D. Constantin Mocanu, R. Barresi, G. Fortino, A. Liotta, Cognitive streaming on android devices, in *Proceedings of the 1st IEEE/IFIP IM 2015 International Workshop on Cognitive Network and Service Management*, Ottawa, Canada, 11–15 May 2015 (IEEE). http://www.cogman.org

17. J. Okyere-Benya, M. Aldiabat, V. Menkovski, G. Exarchakos, A. Liotta, Video quality degradation on IPTV networks, in *Proceedings of International Conference on Computing, Networking and Communications*, Maui, Hawaii, USA, 30 Jan– 2 Feb 2012 (IEEE)

18. G. Exarchakos, V. Menkovski, A. Liotta, Can Skype be used beyond video calling?, in *Proceedings of the 9th International Conference on Advances in Mobile Computing and Multimedia*, Ho Chi Minh City, Vietnam, 5–7 Dec 2011 (ACM)

19. V. Menkovski, G. Exarchakos, A. Cuadra-Sanchez, A. Liotta, Estimations and remedies for quality of experience in multimedia streaming, in *Proceedings of the 3rd International Conference on Advances in Human-oriented and Personalized Mechanisms, Technologies, and Services*, Nice, France, 22–27 Aug 2010 (IEEE)

20. V. Menkovski, G. Exarchakos, A. Liotta, Machine learning approach for quality of experience aware networks, in *Proceedings of Computational Intelligence in Networks and Systems*, Thessaloniki, Greece, 24–26 Nov 2010 (IEEE)

21. F. Agboma, A. Liotta, User-centric assessment of mobile content delivery, in *Proceedings of the 4th International Conference on Advances in Mobile Computing and Multimedia*, Yogyakarta, Indonesia, 4–6 Dec 2006

22. F. Agboma, A. Liotta, Managing the user's quality of experience, in *Proceedings of the second IEEE/IFIP International Workshop on Business-driven IT Management (BDIM 2007)*, Munich, Germany, 21 May 2007 (IEEE)

23. F. Agboma, M. Smy, A. Liotta, QoE analysis of a peer-to-peer television system, in *Proceedings of the International Conference on Telecommunications, Networks and Systems*, Amsterdam, Netherlands, 22–24 July 2008

24. F. Agboma, A. Liotta, QoE-aware QoS management, in *Proceedings of the 6th International Conference on Advances in Mobile Computing and Multimedia*, Linz, Austria, 24–26 Nov 2008

25. V. Menkovski, A. Oredope, A. Liotta, A. Cuadra-Sanchez, Optimized online learning for QoE prediction, in *Proceedings of the 21st Benelux Conference on Artificial Intelligence*, Eindhoven, The Netherlands, 29–30 Oct 2009, pp. 169–176. http://wwwis.win.tue.nl/bnaic2009/proc.html. ISSN: 1568-7805

26. V. Menkovski, A. Oredope, A. Liotta, A. Cuadra-Sanchez, Predicting quality of experience in multimedia streaming, in *Proceedings of the 7th International Conference on Advances in Mobile Computing and Multimedia*, Dec 2009 (ACM). http://dl.acm.org/citation.cfm?id=1821766. ISBN: 978-1-60558-659-5

27. V. Menkovski, G. Exarchakos, A. Cuadra-Sanchez, A. Liotta, Measuring quality of experience on a commercial mobile TV platform, in *Proceedings of the 2nd International Conference on Advances in Multimedia*, Athens, Greece, 13–19 June 2010 (IEEE)

28. V. Menkovski, G. Exarchakos, A. Liotta, Online QoE prediction, in *Proceedings of the 2nd IEEE International Workshop on Quality of Multimedia Experience*, Trondheim, Norway, 21–23 June 2010 (IEEE)

29. V. Menkovski, G. Exarchakos, A. Liotta, Online learning for quality of experience management, in *Proceedings of the Annual Machine Learning Conference of Belgium and The Netherlands*, Leuven, Belgium, 27–28 May 2010. http://dtai.cs.kuleuven.be/events/Benelearn2010/submissions/benelearn2010_submission_20.pdf

30. V. Menkovski, G. Exarchakos, A. Liotta, The value of relative quality in video delivery. J. Mobile Multimedia **7**(3), 151162 (2011). http://dl.acm.org/citation.cfm?id=2230537

31. F. Agboma, A. Liotta, Addressing user expectations in mobile content delivery. Mob. Inf. Syst. **3**(3, 4), 153164 (2007). http://dl.acm.org/citation.cfm?id=1376820.1376823

32. W.H. Ehrenstein, A. Ehrenstein, Psychophysical methods, in *Modern Techniques in Neuroscience Research*, 1999, pp. 1211–1241. http://uni-leipzig.de/~isp/isp/history/texts/PSYPHY-M.PDF

33. V. Menkovski, G. Exarchakos, A. Liotta, Tackling the sheer scale of subjective QoE, in *Proceedings of 7th International ICST Mobile Multimedia Communications Conference*, Cagliari, Italy, 5–7 Sept 2011, vol. 29, pp. 1–15 (Springer, Lecture Notes of ICST, 2012). http://www.springerlink.com/content/p1443m265r25756x/. doi:10.1007/978-3-642-30419-4_1

34. ITU, 500-11, Methodology for the subjective assessment of the quality of television pictures, recommendation ITU-R BT. 500-11, ITU Telecom. Standardization Sector of ITU (2002)

35. F. De Simone, M. Naccari, M. Tagliasacchi, F. Dufaux, S. Tubaro, T. Ebrahimi, Subjective assessment of H.264/AVC video sequences transmitted over a noisy channel, in *International Workshop on Quality of Multimedia Experience, QoMEx 2009*. IEEE 2009, pp. 204–209 (2009)

36. Q. Huynh-Thu, M. Ghanbari, Modelling of spatio-temporal interaction for video quality assessment. Signal Process.: Image Commun. **25**(7), 535–546 (2010)

37. K. Seshadrinathan, R. Soundararajan, A. Bovik, L. Cormack, Study of subjective and objective quality assessment of video. IEEE Trans. Image Process. **19**(6), 1427–1441 (2010)

38. I. Recommendation, 910, Subjective video quality assessment methods for multimedia applications, recommendation ITU-T P. 910, ITU Telecom. Standardization Sector of ITU (1999)

39. M. Barkowsky, M. Pinson, R. Pépion, P. Le Callet, Analysis of freely available subjective dataset for HDTV including coding and transmission distortions, in *Fifth International Workshop on Video Processing and Quality Metrics for Consumer Electronics (VPQM-10)* (2010)

40. S. Stevens, On the psychophysical law. Psychol. Rev. **64**(3), 153 (1957)

41. D. Krantz, A theory of context effects based on cross-context matching. J. Math. Psychol. **5**(1), 1–48 (1968)

42. D. Krantz, A theory of magnitude estimation and cross-modality matching. J. Math. Psychol. **9**(2), 168–199 (1972)

43. D. Krantz, R. Luce, P. Suppes, A. Tversky, *Foundations of Measurement Volume I: Additive and Polynomial Representations*, vol. 1 (Dover Publications, New York, 2006)

44. R. Shepard, On the status of 'direct' psychophysical measurement. Minn. Stud. Philos. Sci. **9**, 441–490 (1978)

45. R. Shepard, Psychological relations and psychophysical scales: on the status of direct psychophysical measurement. J. Math. Psychol. **24**(1), 21–57 (1981)

46. A. Watson, Proposal: measurement of a JND scale for video quality, IEEE G-2.1. 6 Subcommittee on Video Compression Measurements (2000)
47. S. Winkler, On the properties of subjective ratings in video quality experiments, in *International Workshop on Quality of Multimedia Experience, QoMEx 2009*. IEEE 2009, pp. 139–144 (2009)
48. V. Menkovski, A. Liotta, Adaptive psychometric scaling for video quality assessment. Signal Process.: Image Commun. **27**(8), 788–799 (2012)
49. V.Q.E. Group et al., Report on the validation of video quality models for high definition video content, Technical report (2010)
50. P. Corriveau, C. Gojmerac, B. Hughes, L. Stelmach, All subjective scales are not created equal: the effects of context on different scales. Signal Process. **77**(1), 1–9 (1999)
51. P. Brooks, B. Hestnes, User measures of quality of experience: why being objective and quantitative is important. IEEE Network **24**(2), 8–13 (2010)
52. G.T. Fechner, *Elements of Psychophysics* (Holt, Rinehart and Winston, New York, 1966)
53. J.D. McCarthy, M.A. Sasse, D. Miras, Sharp or smooth?: comparing the effects of quantization vs. frame rate for streamed video, in *Proceedings of the SIGCHI Conference on Human Factors in Computing Systems*, ser. CHI'04 (ACM, New York, NY, USA, 2004), p. 535542. http://doi.acm.org/10.1145/985692.985760
54. E.H. Weber, *E.H. Weber On the Tactile Senses* (Psychology Press, Hove, 1996)
55. A. Watson, L. Kreslake, Measurement of visual impairment scales for digital video, in *Human Vision, Visual Processing, and Digital Display. Proceedings*. SPIE 4299 (2001)
56. L. L. Thurstone, *The Measurement of Values*, vol. vii (University of Chicago Press, Oxford, England, 1959)
57. L. Maloney, J. Yang, Maximum likelihood difference scaling. J. Vision **3**(8) (2003)
58. C. Charrier, L. Maloney, H. Cherifi, K. Knoblauch, Maximum likelihood difference scaling of image quality in compression-degraded images. JOSA A **24**(11), 3418–3426 (2007)
59. K. Knoblauch, L. Maloney et al., MLDS: Maximum likelihood difference scaling in R. J. Stat. Softw. **25**(2), 1–26 (2008)
60. D. Green, J. Swets et al., *Signal Detection Theory and Psychophysics*, vol. 1974 (Wiley, New York, 1966)
61. H. Bergsten, *JavaServer Pages*, 3rd edn. (O'Reilly Media Inc., Sebastopol, 2003)
62. JW player: Overview | LongTail video | home of the JW player. http://www.longtailvideo.com/players
63. L. Aimar, L. Merritt, E. Petit, M. Chen, J. Clay, M. Rullgrd, C. Heine, A. Izvorski, VideoLAN—x264, the best H.264/AVC encoder. http://www.videolan.org/developers/x264.html
64. R. Team et al., R: A language and environment for statistical computing (R Foundation Statistical Computing, Vienna, 2008)
65. K. Knoblauch, psyphy: Functions for analyzing psychophysical data in R, R package version 0.0-5. http://CRAN.R-project.org/package=psyphy (2007)
66. B. Efron, R. Tibshirani, *An Introduction to the Bootstrap (Chapman & Hall/CRC Monographs on Statistics and Applied Probability)* (Chapman and Hall/CRC, New York, 1994)
67. F. Wichmann, N. Hill, The psychometric function: I. Fitting, sampling, and goodness of fit. Attention Percept. Psychophys. **63**(8), 1293–1313 (2001)
68. V. Menkovski, G. Exarchakos, A. Liotta, Adaptive testing for video quality assessment, in *Proceedings of Quality of Experience for Multimedia Content Sharing*, Lisbon, Portugal, 29 June 2011 (ACM)
69. S. Kullback, R.A. Leibler, On information and sufficiency. Ann. Math. Stat. **22**(1), 79–86 (1951). http://projecteuclid.org/DPubS?service=UI&version=1.0&verb=Display&handle=euclid.aoms/1177729694
70. K. Eckschlager, K. Danzer, *Information Theory in Analytical Chemistry* (Wiley, New York, 1994)
71. M. Alhaisoni, A. Liotta, M. Ghanbari, Scalable P2P video streaming. Int. J. Bus. Data Commun. Networking **6**(3), 49–65 (2010). doi:10.4018/jbdcn.2010070103. ISSN: 1548-0631
72. M. Alhaisoni, M. Ghanbari, A. Liotta, Localized multistreams for P2P streaming. Int. J. Dig. Multimedia Broadcast. Article ID 843574, 12 pp. Hindawi 2010 (2010). doi:10.1155/2010/843574

73. M. Alhaisoni, A. Liotta, M. Ghanbari, Resource-awareness and trade-off optimization in P2P video streaming. Int. J. Adv. Media Commun. special issue on High-Quality Multimedia Streaming in P2P Environments **4**(1), 59–77 (2010). doi:10.1504/IJAMC.2010.030005. ISSN: 1741-8003

74. A. Liotta, Farewell to deterministic networks, in *Proceedings of the 19th IEEE Symposium on Communications and Vehicular Technology in the Benelux*, Eindhoven, The Netherlands, 16 Nov 2012 (IEEE). http://dx.doi.org/10.1109/SCVT.2012.6399413

Chapter 4
QoE Management Framework

Abstract The focus of this chapter is on monitoring and management of delivered QoE in multimedia systems. Monitoring the QoE typically involves collecting a wide range of available quality performance indicators (QPI) and successfully interpreting these measurements. In this manner, QoE frameworks typically consist of a set of sensors or probes that collect performance data (or QPI) from the system that is fed through QoE models to compute the quality. The QoE estimations delivered by the frameworks indicate the level of performance by the system, against which the service provider can execute the management strategies. This chapter starts with a discussion of existing QoE frameworks and continues to introduce a QoE management framework for an IPTV service developed part of this work (Menkovski et al., Second International Conferences on Advances in Multimedia (MMEDIA), 2010) [1] as a demonstration of an end-to-end QoE managment approach.

4.1 Existing Approaches

There are many proposed solutions for managing the QoE of different service platforms [2–14]. These range from simple monitoring tools that evaluate a restricted set of QPI to complex management systems that correlate many parameters, arbitrate the resources, even adapt the content to the given context [15–34].

The authors of [35] have developed an utility function for each of the following network QoS (NQoS) parameters: delay, jitter, packet loss rate and bandwidth of the video stream. They computed the parameters of a generic utility function based on the results of subjective feedback. The authors further claim that managing the multimedia streams with this utility function approach is more effective than using reservation protocols in today's converged network environments. The proposed approach, however, does not considers the interdependency between the NQoS parameters, but only their effect on QoE independent of each other.

Analysis of quality degradation due to errors during transport is presented in [36]. This methodology focuses on the stream transport statistics to determine the effects of data loss on the video. First, the authors estimate the artifacts in the video due to the transport errors. Then they try to study the visibility of those artifacts and

© Springer International Publishing Switzerland 2015
V. Menkovski, *Computational Inference and Control of Quality in Multimedia Services*, Springer Theses, DOI 10.1007/978-3-319-24792-2_4

their correlation with the perceived quality. The paper discusses a comprehensive analysis of the error handling schemes of H.264 video codec in order to predict the video artifacts. Finally, it continues to analyze the artifacts from the point of view of magnitude (spatial inconsistency and special extent), special priority (region of interest) and temporal duration. The results show that this approach can sometimes follow the trend of MOS results generated by a subjective study better than the PSNR estimations, but the method is still not sufficiently accurate, occasionally even less than PSNR.

Kim et al. present a framework for support of mobile IPTV streaming service in next generation networks (NGN) [37]. The proposed framework uses multiprotocol label switching (MPLS)-based management of the streams to deliver end-to-end QoS guarantees. Redistribution of the available resources is done by considering how much of them are available, what the terminal capability is, and details from the user profiles. Adapting the requested resources for each stream is included in order to optimize the delivered QoE, using technologies such as scalable video coding as well as context-based content extraction. This type of comprehensive adaptations of the content are resource demanding and hard to implement on a real-time system. However, if NGN with the necessary technologies for QoS guarantees are implemented successfully, the framework can deliver improvements in the utilization of the resources to pursue the desired level of QoE.

A QoE modeling and assurance framework also designed for NGN is presented in [38]. In this framework the QoE assurance is guaranteed by the service controller, which intercepts the communication request in the process of establishment. At the point of interception, the controller predicts the level of QoE to be provided within current resources and context. Finally, it adjusts respective quality-related configurations in order to optimize the QoE, given the available resource. This mediation is implemented through the Internet protocol multimedia subsystem that is part of the NGN delivery system [39]. The QoE model is context-aware and uses a comprehensive set of QPI extracted from various information factories in the NGN service delivery system. The model incorporates service and content factors, application factors, and transport factors. However, how these objective factors affect the subjective QoE is not evaluated by this model.

Another framework for delivering QoE aware management in NGN is proposed by Zhang and Ansari [40]. The authors recognize many challenges in delivering end-to-end QoE guarantees, starting from the difficulties in measuring subjective QoE to the fluctuations in resources in wireless networks. They suggest modeling QoE with ML methods; however the solution they propose is restricted to individual parameter thresholds. This approach does not consider the interdependences between different parameters and their joint effect on the QoE. The framework includes of a management and a control block, which determines the target QoE and negotiates with the resource admission control function a way to achieve this QoE. Finally, the authors propose a mechanism that implements a global controlled degradation in QoE when the available resources are not sufficient. However, since QoE is highly non-linear, a better approach might be to refuse new service requests, rather than degrading the QoE of all users.

The following sections presents a QoE monitoring framework that we have developed. This framework is extensible to any number of parameters or QPI, and models the QoE based on subjective feedback using ML methods. Finally, it uses a novel method for calculating possible remedies, which allows for improvement of the QoE per active service or globally. As a case study, this framework is applied to a mobile IPTV system.

4.2 QoE Management for Video Streaming

This section presents a framework for QoE-aware management of a video streaming service. The framework is particularly designed to work in conjunction with an IPTV service for mobile devices. However, the architecture is generic and compatible with many similar multimedia services and can be easily adapted to work with other multimedia content delivery systems.

4.2.1 Architecture of a Video Streaming Systems

The typical video streaming system consists of content servers, transport network and terminal devices (Fig. 4.1). The content servers and the network are commonly referred to as content distribution network (CDN). This allows for the content to be distributed in such a manner that good scalability is archived. The content itself, needs to be encoded and compressed prior to distribution. All these processes need to be efficiently managed in order for the service to be successful.

The management of the video encoding process has a significant impact on the overall efficiency of the system. During encoding, a trade-off between the size of the compressed video and its quality is made. In Chap. 3 we elaborated on how the video quality degrades when the coding bit-rate is restricted. On the other hand, one of the

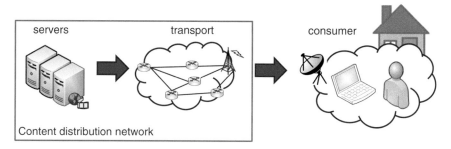

Fig. 4.1 Components of a video streaming system

major costs in multimedia systems is incurred by the storing and transporting of large amounts of video data. Finally, the cost is not the only factor, video streaming delivery requires accurate and timely delivery. Since the network is a shared and limited resource, large video bit-rates encounter severe hurdles. Consequently, managing the encoding process efficiently involves a difficult trade-off decision between resources and quality. Considering the number of factors involved and their non-linear interdependencies, this task becomes a significant challenge.

Video streaming systems may be implemented using a variety of streaming technologies. The content can be encoded in a single layer, multilayer or in multiple levels for adaptive streaming. The multilayer and adaptive technologies offer more control in the degradation of quality when the network resources are insufficient but require more storage and computing resources. Additionally, as these technologies add complexity to the system, they also add more parameters under the management process. Multilayer video has a number of improvement layers in addition to the base layer, with different levels of bit-rate. Similarly, the adaptive streaming video has more than one parallel streams with different bit-rate levels. These additional features improve the performance of the system, but also increase the management complexity.

The transport of the video can be implemented also with different technologies. Progressive download transport is implemented over the HTTP [41] protocol, which runs over TCP [42] in the network. In this case, delivery is guaranteed by the network protocol, so no error correction mechanisms are necessary in the application layers. However, the same TCP mechanisms that guarantee the delivery can produce delays that decrease the transport efficiency. This can result in video data arriving late and causes freezes in the playback. For this reason, progressive download is more suitable for playback of stored content when sizable buffering is feasible. The HTTP adaptive streaming [43], offers more flexibility to the client software in case of insufficient transport resources. The client can choose on-the-fly between different levels of bit-rate, according to the performance of the network, while the TCP network protocol guarantees the accuracy of the data.

RTSP/RTP is an application level protocol dedicated to multimedia streaming [44], which can be implemented on either the TCP or the UDP transport level protocols [42, 45]. Removing the TCP transmission control gives more flexibility to the video streaming protocol to implement the control that is more suited for video transmission. Removing the TCP delays allows for less buffering of the video, while other error handling methods can be implemented such as forward error control (FEC) to handle transmission errors [46].

All these mechanisms provide additional flexibility to the system in dealing with errors or lack of resources, but naturally they also add more complexity to the management process.

With so many factors being part of the management and due to the intricate relationship between the resources and the delivered quality, determining the optimal management strategy becomes a problem that is hard to scale. As more content is added, with different characteristics, and similarly new devices become part of the system, the management effort grows very substantially. In the rest of this chapter we present a set of methods and technologies involved in the implementation of our QoE framework, aimed at dealing with this increasing complexity.

4.2.2 A Hybrid QoE Management Framework

The three high-level components of the video delivery services (Fig. 4.1) are traditionally managed independently of each other. Server resources are managed based on utilization statistics. Similarly, network dimensioning is based on its resource utilization. Since both the servers and the transport provide only best-effort reliability, management usually relies on over-provisioning to keep the service quality high. Content encoding is commonly done in a one-size-fits-all fashion to simplify service management.

This approach leads to a sub-optimal utilization of the available resources. Furthermore, due the fact that the components are managed in a disjoint fashion, information from the transport is not used to optimize the server resource and vice versa. Finally, the video content is not adapted to the terminal devices and to the available transport resources.

In order to improve the management process, a closed loop system needs to be implement, as depicted in Fig. 4.2. The information from the server, transport and terminal device domain need to be correlated with subjective feedback and sent back to the system. In this way, all available measurements can be efficiently used to optimize the management decisions. Based on this approach, we have designed a QoE monitoring and management framework that works in two phases (Fig. 4.3).

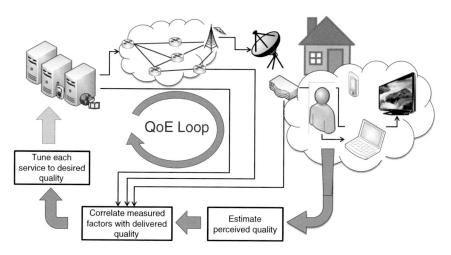

Fig. 4.2 The QoE loop architecture

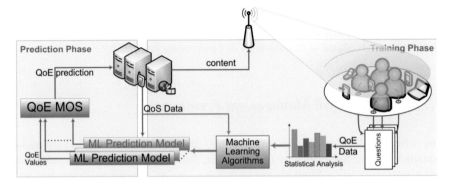

Fig. 4.3 System architecture of the QoE monitoring framework

The first phase is the training phase. For a number of streams, a range of all available QPI and system parameters is measured. Typical available metrics are: packet loss rate; frame rate; video and audio bit-rate; spatial and temporal information; Cp and Cm indexes; video encoding quantization level or quality settings for variable bit-rate; audio sampling frequency and quantization precision, screen size; video resolution; number of playback freezes; and average freeze length. In addition to these objectively collected metrics the platform collected subjective feedback from the viewers of those streams. The subjective feedback is used to model the relationship of the QoE to the values of the measured QPI using ML methods.

The second phase is the operational phase. Using the QoE models, the framework estimates the delivered QoE of each subsequent stream. This way the system performance is monitored continuously. The effect of different management decisions can be observed on the delivered QoE. In order to further enhance the decision process the platform calculates 'remedies' or management decisions, that can effectively improve the delivered QoE, for specific streams or globally for the system.

The second phase is also expanded to include an online training capability that works in conjunction to the monitoring and management. The online learning capability offers continuous adaptation and improvements to the models, as soon as new subjective feedback becomes available. In this way the framework does not need to go back to initial phase when the performance of the models degrades.

The details of the implementation of the QoE monitoring and management framework are presented in the rest of this chapter.

4.3 QoE Management for a Mobile IPTV Service

The Mobile IPTV QoE management framework is proof-of-concept implementation of the hybrid QoE management framework described in the previous section.

This framework works in conjunction with a mobile TV streaming service of a commercial telecom provider [1, 47]. The customers can select to watch one of the available multicast video channels with a fixed quality settings. Some of the content is offered on multiple channels with different quality settings or adapted to specific devices.

Many providers find it more efficient to maintain more than one stream with different qualities rather than a single multilayer stream, due to compatibility issues with end user devices.

In the rest of this section we present the details about the objective and subjective data collected. The Algorithms used for modeling the relationships between the two in an offline and online fashion. Finally, a discussion on the computation of the QoE remedies is given, the last aspect of our QoE management framework for mobile IPTV.

4.3.1 Objective Measurements

In order to monitor the service quality, a probe-based network monitoring system is in place, gathering information from the Mobile TV content distribution platform. For each stream the probes collect information, such as type of device, name of channel, stream, duration of the connection; these are captured in an Internet Protocol Detailed Record (IPDR) format [48].

In addition to this information, using mechanisms from RTSP QoS statistics are collected. Some of these values include the number of packets, packet loss ratio for audio, packet loss ratio for video, average delay, maximum delay, and jitter. The full detailed list of parameters is introduced in [47].

In a nutshell, there is a deployed system collecting the AQoS (application QoS) and NQoS (network QoS) data from the system in real-time [49]. The AQoS involves application-level QoS parameters such as video and audio bit-rate and video frame-rate. The NQoS represents the network QoS parameters such as packet loss, jitter, and delay.

The original IPTV platform gives a good overview of the network conditions, providing useful information for dimensioning the resources and managing the parameters of the content encoding. However, it cannot give any information as to how the service is perceived by the end-user. To acquire the user perception a subjective feedback mechanism was realized as part of this case study, as explained next.

4.3.2 Subjective Measurements

In order to understand the subjective experience of the users, we implemented a subjective study. In this study we ask participants to use the service in various conditions and we collect their feedback on the experience. During the subjective studies, the system records the IPDR values for the specific service provided. Then, each participant fills in a questionnaire. These responses are aligned with the IPDR records correspondingly. After this, the measured objective and subjective values are used as training data for the Machine Learning algorithms. These algorithms produce the models that estimate the QoE are used in the following phase. This approach produces one prediction model for each question. The input to each model are the collected system measurements and the output is a predicted answer to one of the questions. The outputs are later combined to produce an overall QoE value. As long as there is no radical change to the environment (e.g. new device or user group) these models are expected to perform accurate predictions of a subset of QoE values.

The QoE prediction models are plugged in the QoE management framework for online use. In this manner the framework continuously evaluates the performance of the system.

4.4 Computational Inference of QoE Models

When we set out to model the performance of a service, we proceed to design a function that computes the performance from the measured service parameters. We can make one such example using the plain telephony service. Since we know that the human voice is in the range of 300–3400 Hz, the service needs a transmission channel of 4 KHz to cover this range. Furthermore, any delay bellow 200 ms is considered acceptable [50]. So a system using pulse code modulation with a sampling rate of 8 KHz with less the 200 ms delay can be considered to be delivering a service of high quality.

However, when we have to face a system that is of high complexity, and for which we do not fully understand psychophysical perceptual characteristics, developing a model 'directly' can not be done accurately. Furthermore, if the complexity prohibits an exhaustive psychophysical study and the measurements are afflicted with noise, we need to find a different approach to model the performance. Computational intelligence (CI) methods offer a way of modeling complex relationships by observing sample measurements. This is particularly useful in situations where the environment is not fully observable and there is noise in the measurements. With sufficient amount of labeled sample measurements, these methods can be used to produce a model that will map the input data to the labels. In our case the input data is the objectively measurable parameters and the QPI, while the labels are the subjective QoE responses.

This approach also provides for a way to deal with the continuously growing complexity in multimedia systems in a scalable fashion. Since the QoE is not just about QPI thresholds, the interdependencies between the parameters have significant effect on the final outcome. For example, a certain level of video frame rate may be acceptable for 'head and shoulders' type of video, but may not be good enough for an 'action movie'. The resolution of the video and screen size are closely related. Bit-rate, quantization and frame rate are all affecting each other, and so are the spatial and temporal resolution. Finally, the expectations about the service can skew the perception as well. The expectations depend on many factors, such as the service cost or the quality of competitive services. Startup delay and freezes also have an intricate relationship. The more a video is buffered at the beginning, the smaller is the chance for freeze during playback. Many interdependencies between the parameters exists. As they are highly nonlinear, modeling them becomes challenging with traditional statistical inference methods; so adopting modern CI methods is necessary. The following sections details CI algorithms that are suited for this challenge and that were particularly applied in our mobile IPTV QoE management framework.

4.4.1 Supervised Learning Background

Supervised learning algorithms build models based on training data or training examples [51]. Formally, the training set is given in the form of $X = \{\bar{x}^t, y^t\}_{t=0}^N$, where $\bar{x}^t$ is an input vector of attributes and y^t is a class or a label of the tth datapoint in a dataset of size N. For training QoE models, the attributes $\bar{x}$ can be a set of QoS attributes such as video bit-rate, frame rate, and audio bit-rate, while the label is the value of the QoE (good, fair, bad). The goal of the procedure is to derive a hypothesis h about the input data such that $y = h(\bar{x})$. This hypothesis represents our prediction model and can have many forms, such as decision tree, artificial neural net, or a support vector machine.

4.4.1.1 Decision Trees

Decision Trees (DT) are models represented by a hierarchical tree structure where each branching node represents a test (or a question) and each leaf is associated with one possible decision (class or a label) (Fig. 4.4). ML induction tree algorithms

Fig. 4.4 Decision tree

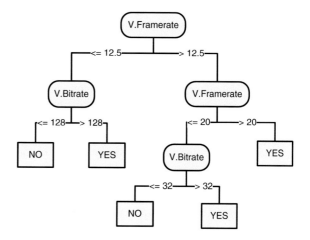

produce DT models from training data in a supervised learning fashion. The DT model, can be used for the classification of unlabeled datapoints. The datapoint values are tested at each branch starting from the root of the tree. The tested datapoint satisfies tests on the route leading only up to a single leaf of the tree. The class (or label) associated to that leaf is the classification output for that datapoint.

The basic concepts for induction of DT are captured by the ID3 algorithm presented in the seminal work by Quinlan [52]. The DT is built by adding branching nodes, starting from the root of the tree. The tree is finalized by adding the edges or the leaves. The tree is built in a recursive fashion until all the branches finish in leaves.

The tests in the branches are expressed as $a_k = v_l$, where a is one of the k attributes of the dataset and v is one of the l possible values of this attribute. If the attribute is of continual nature (real number) than the test is of the form $a_k \geq v$. The attribute a and the value v are selected such that the two subsets (for binary trees) that this test splits the training data into are with minimum entropy in regards to their label. In other words, the dataset is split in such a way that the probability of each datapoint to belong to a single class in each of the subsets is maximized. This is achieved by calculating the information gain for each test.

The information gain $G(S, a)$ of splitting the set S over the attribute a is given in Eq. 4.1.

$$G(S, a) = E(S) - \sum_{i=1}^{m} f_S(a_i) E(S_{a_i}) \tag{4.1}$$

$$E(S) = -\sum_{j=1}^{N} f_s(j) \log_2 f_s(j) \tag{4.2}$$

$E(S)$ is the entropy of the set S calculated as in Eq. 4.2 and $f_S(j)$ is the proportion of datapoints in set S belonging to class j.

Selecting first the tests that split the data into more uniform sets results in a shorter tree that generalizes well. When the subsets only contain datapoints of a single label, the branching is stopped and a leaf associated with that label is attached to that branch.

Using the same principles, the C4.5 algorithm is developed as an extension of ID3 [53]. This algorithm overcomes many weaknesses of ID3, such as handling continuous attributes, training data with missing values, and many different pruning enhancements that deal with overfitting.

We have used the DT methods to build models mapping the many measured parameters of the multimedia delivery system to the subjective quality feedback collected from the users. DT models are easy to use and compute the class of an unlabeled datapoint very efficiently. Furthermore, they represent the model in an intelligible form, which is readable by humans. This is useful for our purpose, since the network operators can derive conclusions about the interdependencies between specific network QoS parameters and the expected QoE.

4.4.1.2 Support Vector Machine

The support vector machine (SVM) is a functional type algorithm and works by first plotting the data in an n-dimensional space (n being the number of attributes). In case of nominal (or discrete) attributes, the algorithm creates another axis for each value of the nominal attribute. One such example is the type of the terminal device. For this attribute the algorithm creates one variable for each possible value. These variables take Boolean values (0 or 1) depending the presence of that particular nominal value.

After plotting the data in the n-dimensional space, the SVM algorithm builds a hyperplane that separates the data in an optimal manner [54] in regards to the two classes (labels). If there are more classes, the SVM generates one hyperplane for each combination of pair of classes. Substituting the values of the attributes, a particular data point can be placed above or below the hyperplane; that is, it belongs to one or the other class. The particular implementation of SVM used here is called Sequential Minimal Optimization [55].

The SVM is an advanced ML algorithm that in many cases outperforms the DT and other algorithms. However, the drawback in using SVM is that is more complex and requires more computing power.

4.4.1.3 Ensemble Methods

Different ML algorithms have different strengths and weaknesses. No one single algorithm is best suited for capturing all various relationships between the data and the labels. Furthermore, training models using the same algorithm on slightly different datasets can lead to different performance in the models because of the different amount of noise they encountered in the data.

Ensemble methods use a technique for combining multiple learners into a group and utilizing their differences to improve the performance of a single classifier. These methods were developed in an attempt to turn 'weak' classifiers (classifiers that perform slightly better than random) into stronger ones [56].

Ensemble methods deploy multiple classifiers trained with different strategies, which combine their predictions into a more accurate group prediction. Their strength is also in improving the generalization capabilities of the standalone classifier [57].

Bagging is an ensemble method of bootstrap aggregation according to which one base classifier of the ensemble is trained on the whole dataset D and the remaining classifiers on a sub sample of D—sampled uniformly with replacement. For classification, the bagging ensemble uses equal weight voting of all the classifiers to output the class of the datapoint.

Boosting is another ensemble learning method. An example to boosting is the AdaBoost algorithm [58]. AdaBoost generates a sequence of base models $h_1, h_2, \ldots, h_k$ using weighted training sets (weighted by $D_1, D_2, \ldots, D_N$, where N is the size of the dataset). To train the first model h_1, the weights are initialized to the values of $1/N$. To train the consecutive models, the algorithms adapt the weights in such a manner that the training is focused more on the datapoints that were misclassified by the previous classifier. The weights of the datapoints that were classified correctly are multiplied by a coefficient β ($\beta < 0$), which may be calculated in different ways depending on the different implementations of AdaBoost. The misclassified samples weights remain unmodified. Finally, the weights are normalized so that they resemble a probability distribution. Models that misclassify more than half of the datapoints are removed from the ensemble.

In our QoE framework, a specific combination of classifiers and ensembles were used to optimize the performance of the system, as discussed next.

4.4.2 QoE Models for Mobile IPTV

Here we present two sets of models for QoE evaluation. Models developed as precursors in the lab and models developed for the commercial IPTV framework case study, respectively.

4.4.2.1 Subjective QoE Models Developed in the Lab

To validate our ML approach, subjective QoE models were initially built in the lab on typical parameters of video streaming services, such as video bit-rate and frame-rate and audio bit-rate. In this study [59] the subjective evaluation is implemented using the 'Method of Limits' [60]. Thus is used to detect the thresholds by changing a single stimulus in successive, discrete steps. A series terminates when the intensity of the stimulus becomes detectable, as described in Chap. 3. For the particular case, we record the segment when the customer has decided that the multimedia quality is unacceptable. The purpose is to determine the user thresholds of acceptability in relation to QoS parameters, taking into account the type of content and terminal. You can see an example of one test for the Mobile terminal in Table 4.1. The user was satisfied with the quality while the video bit-rate was at or above 96 Kbit/s.

Table 4.1 Example of a series of tests in the subjective study

Segment time (s)	Video bitrate (kbit/s)	Audio bitrate (kbit/s)	Frame rate	QoE
1 (1–20)	384	12.2	25	Yes
2 (21–40)	303	12.2	25	Yes
3 (41–60)	243	12.2	20	Yes
4 (61–80)	194	12.2	15	Yes
5 (81–100)	128	12.2	12.5	Yes
6 (101–120)	96	12.2	10	Yes
7 (121–140)	64	12.2	6	No
8 (141–160)	32	12.2	6	No

Fig. 4.5 Decision tree for the mobile dataset

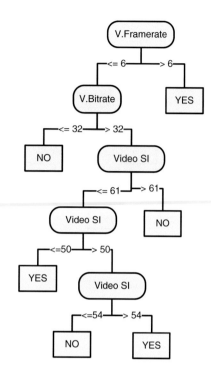

This example generates eight data-points of which six are with a class label of 'Yes' (satisfactory QoE) and two with 'No' (unsatisfactory QoE).

The same tests are performed on different users showing them different video content as well as repeating the tests on three different terminals: mobile, laptop and personal digital assistant (PDA). After compiling the results into three sets for each type of terminal, we used the sets as training data for building prediction models. The J48 algorithm, an implementation of C4.5 [53] in the Weka platform [61], and SMO [55], an implementation of a Support Vector Machine, were used to build the models. The resulting prediction models are shown in Figs. 4.5, 4.6, 4.7, 4.8, 4.9 and 4.10.

The ML models are evaluated for prediction accuracy on a test dataset. The test dataset is usually part of the available labelled data that is reserved for testing and is not used for training. The goal of this estimation is to evaluate how well the model generalizes the concepts found in the data, which could not be accurately accomplished if the same data were used for training and testing.

Test datasets can be generated by reserving approximately 30 % of the available training data for testing. However, when the training data set is of limited size, which is common for subjective data, carving out 30 % of the data for testing is a significant restriction for the training process. In such cases the model evaluation can be implemented using a cross-validation methodology [62]. The 10-fold cross-validation method splits the dataset on 10 equal size sets. Then a model is trained on

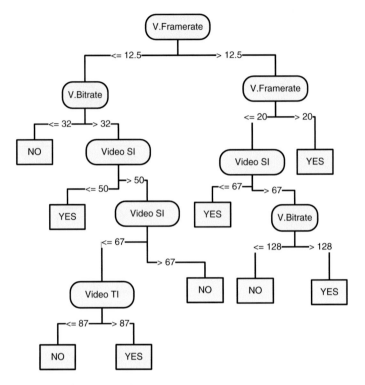

Fig. 4.6 Decision tree for the PDA dataset

9 of the sets and the 10th is used for testing. The procedure is repeated 10 times for each combination of 9 training sets and the one test set. The average performance of the models on the test sets is considered as the performance of the model trained on the whole dataset. The results of the 10-fold crossvalidation for our models are given in Fig. 4.11.

The results from Fig. 4.11 demonstrate that the ML models can be efficiently used for estimating subjective quality.

4.4.2.2 Subjective QoE Models for Mobile IPTV

Using the demonstrated approach we proceed to model the data collected by the QoE framework.

The data for this study is collected by the viewers after watching each of the available videos. The collection is implemented using a Web based interface (Fig. 4.12).

The complete list of questions presented to the participants is the following:

1. What was the type of content that you viewed?
2. What was the amount of delay that you experienced before the video started?

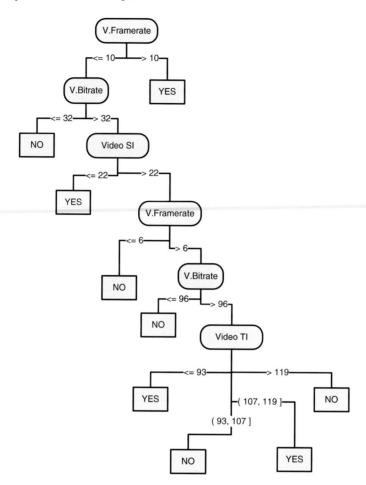

Fig. 4.7 Decision tree for the PDA dataset

Fig. 4.8 SVM hyperplane
for the mobile dataset

```
      1.4555 * (normalized) Video SI
    + 1.0459 * (normalized) Video TI
    - 5.0892 * (normalized) Video Bitrate
    - 3.7632 * (normalized) Video Framerate
    - 0.4582
```

3. Did you experience frozen images or interruptions in the video?
4. Did you experience interruptions in the audio?
5. How much pixelation (big blocks of color) did you experience?
6. Did you experience noise or distortions in the audio?
7. Did you experience problems with the synchronization of the audio and video?
8. How did you find quality of colors?

Fig. 4.9 SVM hyperplane
for the PDA dataset

```
    1.4229 * (normalized) Video SI
-   0.4575 * (normalized) Video TI
-   4.2913 * (normalized) Video Bitrate
-   3.1618 * (normalized) Video Framerate
+   1.3385
```

Fig. 4.10 SVM hyperplane
for the laptop dataset

```
    1.4229 * (normalized) Video SI
-   0.4575 * (normalized) Video TI
-   4.2913 * (normalized) Video Bitrate
-   3.1618 * (normalized) Video Framerate
+   1.3385
```

Fig. 4.11 Performance of
the DT and SVM models on
the three datasets

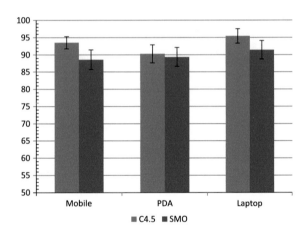

9. How did you find definition (sharpness) of the video?
10. What was your overall perception of the quality?

For the first question the participants could chose from 7 possible answers for the types of content give bellow:

1. News
2. Music Videos
3. Entertainment
4. Documentary
5. Movie or TV Series
6. Cartoon
7. Sports

In questions 2 though 7 the participants are asked to respond between one of the 4 given values bellow:

1. None
2. Little

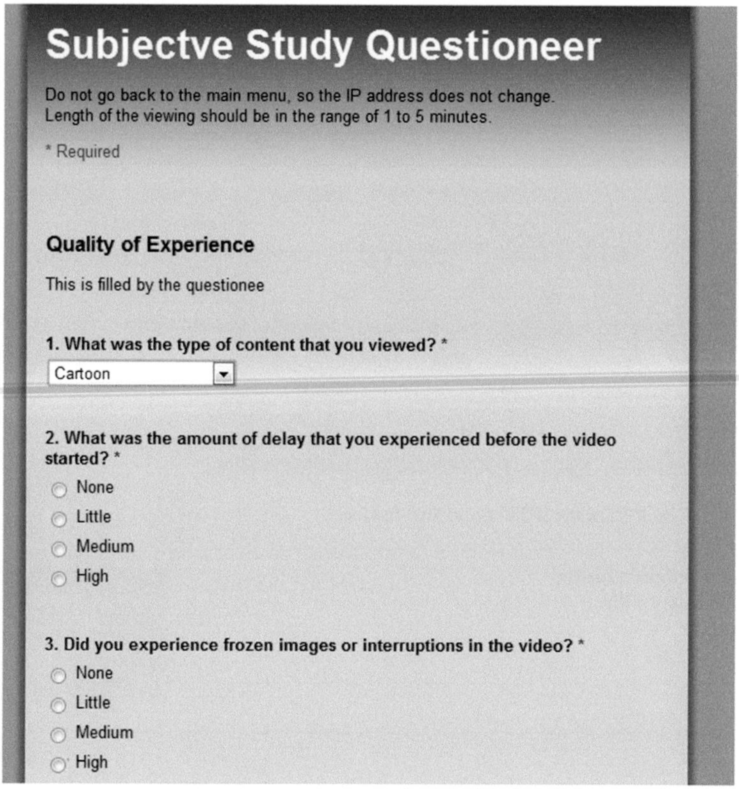

Subjectve Study Questioneer

Do not go back to the main menu, so the IP address does not change.
Length of the viewing should be in the range of 1 to 5 minutes.

* Required

Quality of Experience

This is filled by the questionee

1. What was the type of content that you viewed? *

Cartoon

2. What was the amount of delay that you experienced before the video started? *

None
Little
Medium
High

3. Did you experience frozen images or interruptions in the video? *

None
Little
Medium
High

Fig. 4.12 Subjective study questionnaire

3. Medium
4. High

In questions 8 and 9 the response is between the following 4 values:

1. Excellent
2. Acceptable
3. Poor
4. Unacceptable

The final 10th question offered a choice of 5 distinct values:

1. Excellent
2. Very good
3. Not so good
4. Very bad

The collected subjective dataset contains 55 features from the IPDR log files and the 10 subjective responses from the viewers. For each of the 10 subjective responses

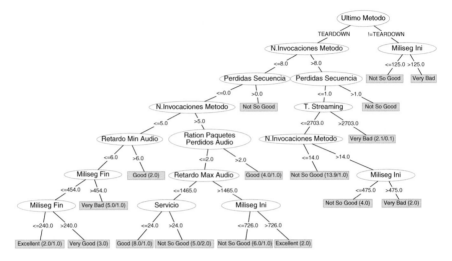

Fig. 4.13 DT built from the IPTV subjective feedback

Fig. 4.14 Confusion matrix
for high accuracy with
tolerance ±1

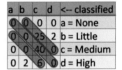

a	b	c	d	<-- classified
0	0	0	0	a = None
0	0	25	2	b = Little
0	0	40	0	c = Medium
0	2	6	0	d = High

a separate model is trained. Finally a combination of the output of the 10 models is used to predict the overall QoE.

Each of the models is trained using C4.5 and SVM as base learners of an AdaBoost ensemble. The performance of the QoE models is given if Fig. 4.13.

The models were trained using categorical labels, which make them easy for humans to read and understand, for example "Excellent" or "Not Good". But from the prediction point of view these labels are not ordered, they are considered the same as we would consider the labels "Red" or "Blue". So when we are calculating the prediction accuracy, only the exact predictions are taken into account as accurate. We do not know how many near misses we have. Most of these near misses would provide for good management input. For instance, a prediction of "Very Bad" and "Not Good" might lead to the same management decision since both cases are not satisfactory. If we take this into account and tolerate a small error rate for the output, such as errors with a distance of one or less from the actual value, the accuracy of the models significantly increases. For graphical representation we can look at the confusion matrix in Fig. 4.14; the main diagonal represents the accurate cases (actual value row and predicted value column). If we add the values in the two adjacent diagonals, we can get the new accuracy with tolerance of ±1 and thus get a higher effective accuracy of our classifier (Fig. 4.15).

Fig. 4.15 Prediction
accuracy with and without
output aggregation

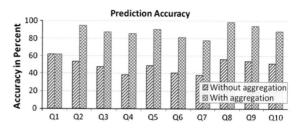

4.5 On-line Inference of QoE Models

Supervised learning models give us the possibility to model the QoE from subjective data. However, multimedia systems are evolving rapidly, with the introduction of new services and new devices. The models trained on the pre-determined conditions become less accurate as conditions change. Retraining the models with new subjective studies regularly is costly and inefficient. Instead we propose the use of an online learning approach, where subjective feedback is received continuously and is used for updating the models with the new data. In order for this method to be incorporated, we need to have a mechanism for continuous collection of subjective feedback and new set of methods for inference of models that work in on-line fashion.

The rest of this section introduces the on-line supervised learning methods used in our QoE management framework.

4.5.1 Online Supervised Learning Background

The online learning algorithms are ML algorithms designed to train models in a supervised learning fashion from labelled data. The only difference with offline or batch learning is that the data is processed sequentially and continuously. The main motivation for online learning is modelling fast data streams without having to retain a sizable amount of data. In order to achieve this, the algorithms need to be able to add new concepts to their models online, i.e. as new data becomes available. Moreover, they also need to be able to 'forget' or remove concepts from the models that are not present in the incoming data.

Changes in perceived QoE commonly occur when new types of content or new services are introduced, but also when the user's expectations increase with the advances in technology. This kind of change can be quite frequent in multimedia services. To circumvent the need to redo the subjective studies and recreate the models, we included online learning technology in the framework.

In the following sections we detail the online learning methods used in the framework, providing an analysis of their performance.

4.5.1.1 Hoeffding Trees

Hoeffding trees (HT) [63] is an algorithm for decision tree induction that is designed to handle extremely large training sets delivered by fast data streams. The training set is commonly so large that it is not expected for the training data to remain in memory. It is actually processed from the input stream in a single pass. The fact that the data is processed sequentially, or one data-point at the time, characterizes this approach as online learning.

At any point in time, the learner has only a partial view of the data, since the rest of the dataset has not yet been introduced. This means that the selected attribute for the test in a node cannot be made with full confidence for any split criteria, but it has to be made with a more relaxed one. The algorithm selects the best attribute at a given node, by considering only a small subset of the training examples that satisfy all the tests leading to that node. As the data is being introduced, the first datapoints are used to choose the root test. Once the root attribute is selected, the succeeding examples will be passed down to the corresponding leaves and used to choose the appropriate attributes for the tests in the new branching nodes that replace existing leaves, and so on. The number of examples that justify a branching at each node is made by relying on a statistical result known as Hoeffding bound. Given n observations of a random variable r with a range R, the calculated mean of r is $\hat{r}$. The Hoeffding bound states with probability $1 - \delta$ that the true mean of the variable is $\hat{r} - \epsilon$ whereby ϵ is given in Eq. 4.3

$$\epsilon = \sqrt{\frac{R^2 ln(1/\delta)}{2n}} \qquad (4.3)$$

Defining the attribute selection criterion as $G(a)$, then $\Delta G = G(a_1) - G(a_2) > 0$, assuming that the a_1 attribute is more favourable (or with larger information gain) than a_2. Given the desired ϵ, the Hoeffding bound guarantees that a_1 is the better selection with probability δ if n examples are seen, where $\Delta G > \epsilon^2$.

In addition to the relaxed information gain criteria for generating the branches, the HT algorithm also implements mechanisms for pruning existing branches. At each branch, attribute selection is also being tested against a'dummy' test on the attribute a_0 that substitutes the branch with a leaf. If this test results in better gain for the sufficient number of tests given by the Hoeffding bound, then this branch is pruned.

The resulting tree constantly adapts as new data is being introduced, capturing more and more relationships between the attribute values and the labels. Moreover, it also deletes relationships that disappear from the newly introduced datapoints as new trends in the data appear. This makes this method applicable for building QoE models based on continuously collected subjective feedback.

Two further incremental improvements to the HT algorithm that are used in the application to our framework are explained next.

Hoeffing option trees
Option trees generalize the regular decision trees by adding a new type of node, an option node [64]. Option nodes allow several branchings instead of a single branching per node. This effectively means that multiple paths are followed below the option node and classification is commonly done by a majority-voting scheme of the different paths. Option Decision Trees can reduce the error rate of Decision Trees by combining multiple models and combining predictions while still maintaining a single compact classifier. For the proposed methodology, the combination of the predictions of different paths is done with weighted voting [65], which sums up individual probability predictions of each leaf.

Hoeffding Option Trees with functional leaves
The usual way a decision tree is built by assigning a fixed class to each leaf during the training. This class is equal to the class of the majority of the training datapoints that reach this node. There is another approach based on which the leaves are not associated with a fixed class but are simple classifiers themselves (referred to as functional leaves). These classifiers are trained only on the data that falls on the leaf during the training of the whole tree. This approach can outperform both a standalone decision tree as well as a standalone classifier [66]. In further research on functional leaves the authors of [67] show that, for incremental learning models, naïve Bayes classifiers used as functional leaves improve the accuracy over the majority class approach. However, this cannot be a rule of thumb. There are exceptional cases shown in [68], where a standard Hoeffding Option Tree will outperform the tree with functional nodes. The author of [68] proposes an adaptive approach, where the training algorithm adaptively decides to use the functional or majority votes, based on the current performance of each of them. This implementation is adopted in our framework to enable efficient online learning.

4.5.1.2 Online Ensemble Methods

Earlier in this chapter we have discussed the benefits of using ensemble methods for improving the performance of supervised learning algorithms. However, modifications for the ensemble techniques are also necessary to enable the online mode of operation.

First of all, the online ensemble methods need to have online algorithms as base learners, so that they can update their base models as new data arrives. However, in batch or offline learning the ensemble algorithm has the freedom to split the set in different subsets or change the dataset weight distribution for each learner. On the other hand, in online learning the data arrives one at the time, so the online ensemble algorithms need to adapt their strategy accordingly.

In online bagging [69, 70], instead of resampling the data with replacement as in offline bagging, the algorithm presents the datapoint $(\bar{x}, y)$ to each learner multiple times. The number of presentations of a datapoint to a base learner is K times, where K is sampled from a *Poisson*(1) distribution. The authors of [70] claim that the online bagging classifier converge to the normal bagging classifier performance as the number of training examples tend grow to infinity.

Online boosting [69] is designed to be an online version of AdaBoost [56] algorithm. As we described previously AdaBoost generates a sequence of base models $h_1, h_2, \ldots, h_k$ such that each consecutive model's training is focused on the datapoints misclassified by previous models. The online version of this algorithm repeats the presentation of the datapoints similarly to the online version of bagging. The only difference is that the number of presentations K to a based learner t is sampled from a distribution $Poisson(\lambda)$, where the parameter λ is increased if the previous model (h_{t-1}) misclassified the datapoint or decreased if h_{t-1} classified the datapoint correctly.

4.5.2 Subjective QoE Models with Continuous Learning

To evaluate the online learning approach, we implement a set of experiments where subjective data is introduced to the learners sequentially in an online fashion and their performance is evaluated. The subjective data is collected with the method of limits [71] and have two labels, 'acceptable' and'not acceptable'. We consider different kinds of streamed content on three different terminals, as described in the previous chapters.

For the implementation of the online learning algorithms we used the Massive Online Analysis (MOA) [72] ML platform for data stream mining, which has implementations of Hoeffding Option Tees and Oza Bagging algorithm. The MOA platform is an extension of the WEKA ML data mining platform [61]. In our case, the viewer feedback is considered as scarce and expensive, so we can only expect small amount of data arriving. In light of this issue we have modified some of the parameters of the algorithm to serve our purpose, mainly the n_{min} grace period from 200 (default value) to 1. It is not meaningful to wait for 200 datapoints until we start building the DT when we only have 3500 datapoints available.

An adaptation of the model evaluation procedure is also necessary. In MOA there is the assumption of abundance of data, and the estimation of the accuracy of the prediction models is done by interleaving testing and training. In this way there is part of the data that is dedicated for testing, and this data is not used for training of the models. Consequently, the accuracy of these models could be reported as lower in cases of small amount of available data. The approach for model evaluation in cases of scarce data is cross-validation as described in the previous section.

We implemented a 10-fold cross validation scheme to calculate the accuracy of the classifier. This validation scheme splits the data into 90% for training and 10% for testing, and then it repeats this process ten times. Each time different combination of datapoints is used for training and testing.

The results of the execution of the Online Learning using the Hoeffding Option Tree algorithm that the classification accuracy rises fast to over 80% with fewer than 100 datapoints, i.e. user-generated feedback instances. After around 1000 datapoints the classifiers converges to its accuracy of approximately 90% (Fig. 4.16). In the same manner the standard deviation of the accuracy falls quickly to below 3% (with just a few exceptions) and then falls to around 2% after introducing 1600 points.

Fig. 4.16 Hoeffding option tree results

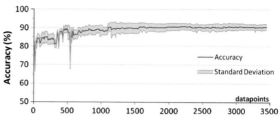

Fig. 4.17 OzaBagging hoeffding option tree results

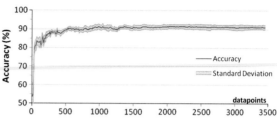

We obtained qualitatively similar findings from the execution of the ensemble Online Learning algorithm Oza bagging Hoeffding Option Tree (Fig. 4.17). We can see that both algorithms reach very high prediction accuracy (90 %) very rapidly (order of a thousand of datapoints). As expected, the ensemble approach gains accuracy faster. Furthermore the classifier's standard deviation of accuracy over the different folds of the cross-validation is much lower than in the stand-alone classifier.

Overall, we have presented results that show that we can already achieve an accuracy of over 80 % by learning only 100 datapoints which are randomly selected feedback.

In order to evaluate the overall benefit of using online learning to handle changes in the environment, we developed the experiment for testing the concept drift. Changes in trends of the labelled data are referred to as concept drift. Classifiers need to adapt to concept drift by modifying or deleting existing concept in the models to accommodate the new patterns in the data. The test is implemented by training the classifier on one type of data from the subjective study and then introducing a new type. At the moment when the new data is introduced the model is not aware of the change and predicts based on knowledge from the previous data. Then new data is introduced from the second dataset. The algorithm updates the model according to the introduced data.

In our algorithm stack, different algorithms behave differently. The Hoeffding Option Tree discards some nodes and induces new ones. The weights on the different paths of the option tree might be modified, and the online ensemble classifier can decide to update the weights to the individual classifiers if their accuracy decreases. With the proposed experimental setup we can monitor the introduction of new concepts in the dataset and the speed of adaptation to the model. The results of the concept drift experiment are presented in Fig. 4.18 for a single HOT classifier and in Fig. 4.19 for an OzaBagging ensemble of HOT classifiers.

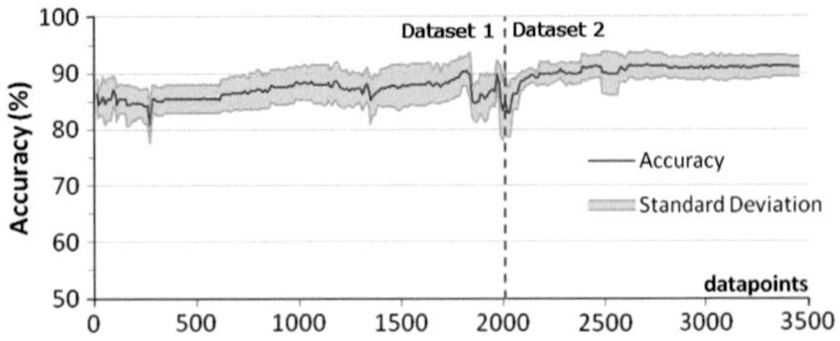

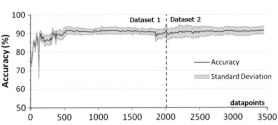

Fig. 4.18 Results of the hoeffding option tree with concept drift experiment

Fig. 4.19 OzaBagging hoeffding option tree with concept drift results

The first dataset contains only video with Temporal Information smaller than 110, which includes about 60 % (2010 out of 3370) of the data. The second set contains the remaining 40 % of data. This are samples with Temporal Information higher than 110, typically content with higher dynamics. The result from using the Hoeffding Option Tree algorithm shows drop in the accuracy and increases in the standard deviation at the moment when the new dataset is introduced. However, the accuracy is recovered very fast and converges to above 90 % in fewer than 200 datapoints. This is a very encouraging result that shows the capabilities of this algorithm to adapt to changes. In this experiment the model was trained on content with small TI (slow changing content) and then we introduced high changing content (high TI). Even with this rather drastic change the accuracy recovered very fast.

The results obtained with the Oza Bagging Hoeffding Option Tree ensemble are even better (Fig. 4.19). This algorithm is much more robust to changes and deals with the concept drift with close-to-none loss in accuracy and limitted rise in the standard deviation. This result shows the tremendous value of online learning, the robustness of the ensemble approach and justifies the added complexity in using an ensemble versus a standalone classifier.

To demonstrate the statistical significance and viability of our approach, Fig. 4.20 illustrates how Hoeffding Option Tree (HOT) and Oza Bagging HOT (OzzaBagg HOT) compare with 3 standard ML approaches, namely Naïve Bayes, Support Vector Machine (SVM) and C4.5. Since the offline algorithms have the advantage to learn on the whole dataset, the performance of the online algorithms is expected as good as the offline in the best case. It is evident, that C4.5 performs best with 93 % accuracy but the online learning algorithms we used follow closely behind with 90.5 and 91.1 %.

We demonstrate the usefulness of this approach by testing it on data that was previously derived via conventional subjective studies. The QoE prediction models show high accuracy and high adaptability to concept drift in the dataset. The fact

Fig. 4.20 Comparison with standard ML algorithms

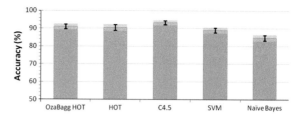

that the accuracy of the online learning algorithms are approaching the accuracy of the standard batch ML algorithms (of above 90 %) demonstrates the applicability of the approach. Unfortunately, due to project constraints and limited access to the commercial IPTV platform, we could not evaluate our online learning system on a real deployment.

4.6 QoE Remedies

In addition to accurate measurement of QoE, efficient management of multimedia services also necessitates efficient provisioning of resources towards the delivered quality. However, as complex relationships exist between the parameters that govern these resources and the delivered quality, selecting the appropriate values is not trivial.

When the measured QoE is unsatisfactory, how do we determine the optimal way to improve it? On the other hand, when the delivered QoE is high, are we over-provisioning certain resources? Can we achieve the same QoE with fewer resources and utilize them more efficiently?

To answer these questions, we need to expand the functionalities of the QoE management framework to include mechanisms that allow for calculating the needed changes, which will provide the targeted QoE. We define these mechanisms as 'QoE remedies' and we proceed to describe them in the rest of this section.

4.6.1 Estimating the QoE Remedies

The QoE remedies are changes to parameters under management that need to be applied to a specific instance of the service so that its QoE is improved. These changes (or distance in values) can be on a single parameter, such as the bit-rate of the video. Or they can be on a combination of multiple parameters, such as frame-rate and the resolution of the video. The remedies can also refer to measurements such as the incurred packet loss. However, decreasing the packet loss may not be directly under the control of a single parameter and may include the utilization of multiple resources. For example, the level of packet loss can be affected by

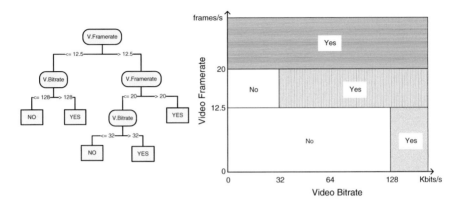

Fig. 4.21 Simple decision tree in 2D space

increasing the network resources or by using more advanced transmission data protection schemes. In any case, the QoE remedies deliver a deeper level of understanding of the correlation between the system resources and the resulting service QoE. This level of understanding enables more informed management of the service and alleviates the need for a trial and error approach. Furthermore, if the suggested remedies can be associated with costs, the management system can autonomously select the best option. In this manner a high level of autonomic behaviour can achieved as the framework will take over many of the monitoring and management processes.

To accomplish this task we implemented an algorithm that, based on the QoE prediction model, estimates the minimum needed changes in the measured stream parameters to improve the QoE [73]. This technique is enabled by the DT prediction models we use for estimating the QoE. The algorithm represents a QoE prediction DT model in the geometric space, defined by the dataset parameters. For each parameter it defines a single dimension in the parameter hyperspace. Each of the datapoints from the dataset can be represented as a point in this hyperspace. The DT is finally represented by hyper-regions formed by the leaves of the DT (Fig. 4.21).

Each node of the DT represents a binary split (for binary trees) that forms a hyperplane in the parameter hyperspace. At the bottom, the leaves of the tree carve out hyper-regions, which are bounded by these hyperplanes. These hyper regions are associated with a class label membership, according to the leaf they correspond to. Every datapoint in the dataset actualizes the tests of a route leading to only a single leaf on the DT. Correspondingly, every datapoint falls within a single hyper-region and is therefore classified with the label for that region.

Our algorithm represents the DT in the hyperspace by generating a set of hyper regions that correspond to the tree leaves (Fig. 4.22). Each hyper-region contains a set of split rules that define the hyper-surface, which define the boundaries of the hyper-region. The split rules are either representing an inequality of the type $Parameter_1 \geq Value_1$ or of the type $Parameter_1 = Value_1$ depending on whether $Parameter_1$ is continual or categorical. If the leaf is on the left side of a continual $Parameter_1$ split then the split inequality will be '*more than or equal to*', if it is on the right side

Fig. 4.22 Hyper-region
building algorithm

> Start from the root node and call a recursive method
> *FindLeaves*
>
> *FindLeaves:*
>
> 1) If the node has children
>   a) Call *FindLeaves* on each child
>   b) Add the SplitRule on each of the Hyper Regions ($\overline{\Phi}$)
>   that are returned
>   i)   If the leaf split is categorical add a Split Rule:
>   Attribute = 'value'
>   ii) If the leaf split is continual add on the leaves
>   from the left side SplitRule: Attribute < value,
>   and on the leaves from the right side Attribute >
>   value
>   c) Return the set of Hyper Regions ($\overline{\Phi}$)
> 2) Else, you are in a leaf
>   a) Create an Hyper Region object
>   i)   Assign the class of the leaf to the Φ
>   ii) Return Φ

the split inequality will be '*less than*'. Having a list of *HyperRegion*-s we can easily determine where each datapoint from the dataset belongs to, by testing the datapoint on the split rules of each hyper region. The hyper region is associated with the same class label as the leaf it represents, so all datapoints that belong to that region are classified according to this label.

In order to improve the QoE of a particular instance of the service, we need to look at the measured value that was acquired by the monitoring system for that instance. If the measurement is classified with a QoE value that is not satisfactory, we look at the distance to the hyper regions that are associated with a satisfactory QoE value. The distance to each of the desired regions is the difference in parameter values that are needed in order to move the datapoint to the desired regions. The output of the algorithm is a set of distance vectors, which define the parameters that need to be changed and their change values.

To illustrate the matter better we can take an example from the laptop dataset from [74]. The prediction model built from this dataset is given in Fig. 4.21. If we look at the datapoint given in Table 4.2 we can see that this datapoint will be classified by the model as $QoE = No$ ('*Not Acceptable*'). Since the $V.Framerate$ is <12.5 and the $V.Bitrate$ is <32 the datapoint reaches a leaf with '*Not Acceptable*' class associated with it. Now, what is the best way to improve the QoE of this stream?

First of all there are parameters that can be only observed, such as $VideoSI$ and $VideoTI$ that characterize the type of the content. In this dataset structure we are looking into increasing the $V.Bitrate$ and $V.Framerate$. If we increase the

Table 4.2 Example datapoint

Video SI	Video TI	V.Bitrate	V.Framerate
67	70	32	10

V.Bitrate for this particular datapoint by one step to 64 kbits/s we can see that the datapoint now arrives at one of the bottom leaves of the DT, but it is still classified as *QoE Acceptable = No*. On another hand, if we increase the *V.Framerate* to 15 fps we can see that the datapoint is classified as *QoE Acceptable = Yes* without adding more bandwidth.

We can deduce a rule from the model that a video with these characteristics needs to have higher *V.Framerate* for it to be perceived with high quality. However, this rule is not easily evident from only looking at the model. We can also imagine a system with large number of parameters where changing one or more parameters affects the QoE in a complex way. Further down this line of reasoning, if we want to make a system-wise improvement that will increase the QoE of most streams we cannot easily derive which parameters are best to be increased and by how much.

In the case of the example datapoint the algorithm returns the two possible paths:

- Increasing the Framerate to above 12.5 f/s;
- Increasing the V. Bitrate to above 32 kbits/s and the Video TI to above 87.

Since we know that increasing the *VideoTI* is not an option, because this is a measurement of an inherent characteristic of the video, the only option available is to increase the frame rate. In a general case, there can be many different paths to a hyper-region with the desired class. To automate the process, we can assign cost functions to the change of the attribute values and automatically calculate the cheapest way to reach the desired QoE. In this manner attributes that are observations and cannot be controlled, such as the *VideoTI*, can be excluded from the evaluation by giving them infinitely large cost.

Given a datapoint and a target label, the algorithm produces a set of change vectors. Each of the change vectors applied to the datapoint moves the datapoint to a hyper-region classified with the target label. In other words, each change vector is one possible fix for the datapoint (Fig. 4.23).

$$\bar{\Phi} = FindLeaves(DT, QoE) \qquad (4.4)$$

$$\Delta\varphi_i = Distance(\bar{\Phi}_i, \bar{d}) \qquad (4.5)$$

$$\Delta\varphi_{optimum} = min_i(Cost(\Delta\varphi_i)) \qquad (4.6)$$

In Eq. 4.4, the function returns a set of regions with a targeted QoE value. The distance function in (4.5) calculates the vector of distances for each attribute to the target region in $\Delta\varphi$. The optimal distance vector is the one with minimal cost (4.6) for the given input datapoint $\bar{d}$. The Cost function in (4.6) is dependent on the

Fig. 4.23 Change vector remedies for an example instance

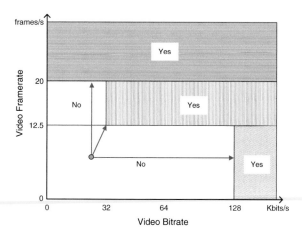

application. Each system has explicit and implicit costs associated with changes of specific parameters.

4.6.2 QoE Remedies for Mobile IPTV

The remedies algorithm has been implemented by extending the Weka [61] platform, so that algorithms such as J48 [53] that induce decision trees can be used to calculate the hyper-regions. Furthermore, we can now measure the distance of any datapoint classified by the DT to the desired hyper-region.

The boxed nodes represent the leaves and map to the hyper-regions as we have seen in Fig. 4.23. There are 17 hyper-regions, out of which, only two are with excellent QoE value. The algorithm generates the remedy output specific for each particular broadcasting system.

A target QoE values needs to be defined, and a specific cost for changing a parameter needs to be given as well. If the target value is 'excellent' QoE, the algorithm will calculate the minimum cost of changing specific parameters so that the datapoint falls in one of the two 'excellent' hyper-regions.

A more elaborate QoE improvement is also possible where not all datapoints are targeted for the excellent regions, but the management is executed based on the utility of improving a QoE of a stream in regards to the costs. Then multiple levels of remedies can be suggested by the algorithm with varying costs, and the provider can chose to apply mechanisms to implement the remedies based on their utility to the customers.

This methodology presents a pragmatic solution for estimating and maintaining QoE with a wide range of applicability. Its success and usability depends on the quality of the prediction models, while as architecture it is flexible enough to be used in many different environments.

Since Online Learning techniques also generate DT models they are compatible with the QoE remedies. Both technologies working together deliver flexible remedy response adapting to the changes in the environment reported by the user QoE feedback.

4.7 Conclusions

In this chapter an approach for QoE enabled management is presented in the form of a framework. The framework correlates signals coming from the monitoring system with subjective feedback from the users to develop QoE estimates and management decisions. A description of its application to a mobile IPTV service is also given.

The subjective data used for developing the QoE models in this chapter relied mostly on rating and method of limits. However, as described in Chap. 3, 2AFC methods show superior performance for collecting subjective data. This framework can further benefit from incorporating this kind of subjective data into the QoE models. Certain challenges still remain for the future work. For example, combining MLDS utility curves for certain QPI with QoE models from rating feedback can improve the management decisions due to the accuracy of these curves. Further modifications to the online learning and the remedy algorithms will also be necessary for fully integrating the MLDS models.

This management framework is designed to address management of resources in multimedia systems reactively, taking into account measurements received from the monitoring system. However, a different set of challenges are faced when control decisions need to be implemented in a proactive fashion. In the following chapter we discuss proactive control in multimedia services, solutions we have developed and the CI technologies that enable these solutions.

References

1. V. Menkovski, G. Exarchakos, A. Liotta, A. Sánchez, Measuring quality of experience on a commercial mobile tv platform, in *Second International Conferences on Advances in Multimedia (MMEDIA)* (IEEE, 2010), pp. 33–38
2. V. Menkovski, G. Exarchakos, A. Liotta, A. Cuadra Sánchez, Quality of experience models for multimedia streaming. Int. J. Mob. Comput. Multimedia Commun. 2(4), 1–20 (2010). www.igi-global.com/ijmcmc/. doi:10.4018/jmcmc.2010100101. ISSN:1937-9412
3. V. Menkovski, G. Exarchakos, A. Liotta, The value of relative quality in video delivery, J. Mob. Multimedia 7(3), 151–162 (Rinton Press, 2011). ISSN:1550-4646, http://www.rintonpress.com/xjmm7/jmm-7-3/151-162.pdf
4. F. Agboma, A. Liotta, Quality of experience management in mobile content delivery systems. J. Telecommun. Syst. (special issue on the Quality of Experience issues in Multimedia Provision) **49**(1), 85–98 (Springer, 2012). doi:10.1007/s11235-010-9355-6

5. V. Menkovski, A. Liotta, Adaptive psychometric scaling for video quality assessment. J. Sign. Process.: Image Commun. **26**(8), 788–799 (Elsevier, 2012). http://dx.doi.org/10.1016/j.image.2012.01.004

6. G. Exarchakos, L. Druda, V. Menkovski, P. Bellavista, A. Liotta, Skype resilience to high motion videos. Int. J. Wavelets, Multiresolut. Inf. Process. **11**(3), (World Scientific Publishing, 2013). http://dx.doi.org/10.1142/S021969131350029X

7. A. Liotta, The cognitive net is coming. IEEE Spectrum **50**(8), 26–31 (IEEE, 2013). http://dx.doi.org/10.1109/MSPEC.2013.6565557

8. G. Exarchakos, V. Menkovski, L. Druda, A. Liotta, Network analysis on Skype end-to-end video quality. Int. J. Pervasive Comput. Commun. **11**(1) (Emerald, 2015). http://www.emeraldinsight.com/doi/abs/10.1108/IJPCC-08-2014-0044

9. C. Ragusa, A. Liotta, G. Pavlou, An adaptive clustering approach for the management of dynamic systems, IEEE J. Sel. Areas Commun. (JSAC). (special issue on Autonomic Communication Systems) **23**(12), 2223–2235 (IEEE, 2005). (IF 4.138; SJR 3.34) http://dx.doi.org/10.1109/JSAC.2005.857203

10. M. Alhaisoni, A. Liotta, Characterization of signalling and traffic in joost. J. P2P Networking Appl. (special issue on Modelling and Applications of Computational P2P) **2**, 75–83 (Springer, 2009). doi:10.1007/s12083-008-0015-5, ISSN:1936-6450

11. M. Alhaisoni, A. Liotta, M. Ghanbari, Resource-awareness and trade-off optimization in P2P video streaming. Int. J. Adv. Media Commun. (special issue on High-Quality Multimedia Streaming in P2P Environments) **4**(1), pp. 59–77 (Inderscience Publishers, 2010). doi:10.1504/IJAMC.2010.030005, ISSN:1741-8003

12. M. Alhaisoni, M. Ghanbari, A. Liotta, Localized multistreams for P2P streaming. Int. J. Digit. Multimedia Broadcast. **843574**, 12 pp. (Hindawi, 2010). doi:10.1155/2010/843574

13. M. Alhaisoni, A. Liotta, M. Ghanbari, Scalable P2P video streaming. Int. J. Bus. Data Commun. Networking **6**(3), 49–65 (IGI Global, 2010). doi:10.4018/jbdcn.2010070103, ISSN:1548-0631

14. A. Liotta, Farewell to deterministic networks, in *Proceedings of the 19th IEEE Symposium on Communications and Vehicular Technology in the Benelux*, Eindhoven, The Netherlands, 16 Nov 2012 (IEEE). http://dx.doi.org/10.1109/SCVT.2012.6399413

15. M. Torres Vega, E. Giordano, D.C. Mocanu, D. Tjondronegoro, A. Liotta, Cognitive no-reference video quality assessment for mobile streaming services, in *Proceedings of the 7th International Workshop on Quality of Multimedia Experience*, Messinia, Greece, 26–29 May 2015 (IEEE). http://www.qomex.org

16. M. Torres Vega, D. Constantin Mocanu, R. Barresi, G. Fortino, A. Liotta, Cognitive streaming on android devices, in *Proceedings of the 1st IEEE/IFIP IM 2015 International Workshop on Cognitive Network and Service Management*, Ottawa, Canada, 11–15 May, 2015 (IEEE). http://www.cogman.org

17. D. Constantin Mocanu, G. Exarchakos, H.B. Ammar, A. Liotta, Reduced reference image quality assessment via boltzmann machines, in *Proceedings of the 3rd IEEE/IFIP IM 2015 International Workshop on Quality of Experience Centric Management*, Ottawa, Canada, 11–15 May 2015 (IEEE)

18. D. Constantin Mocanu, G. Santandrea, W. Cerroni, F. Callegati, A. Liotta, Network performance assessment with quality of experience benchmarks, in *Proceedings of the 10th International Conference on Network and Service Management*, Rio de Janeiro, Brazil, 17–21 Nov 2014 (IEEE)

19. M. Torres Vega, S. Zou, D. Constantin Mocanu, E. Tangdiongga, A.M.J. Koonen, A. Liotta, End-to-End performance evaluation in high-speed wireless networks, in *Proceedings of the 10th International Conference on Network and Service Management*, Rio de Janeiro, Brazil, 17–21 Nov 2014 (IEEE)

20. D.C. Mocanu, A. Liotta, A. Ricci, M. Torres Vega, G. Exarchakos, When does lower bitrate give higher quality in modern video services?, in *Proceedings of the 2nd IEEE/IFIP International Workshop on Quality of Experience Centric Management*, Krakow, Poland, 9 May 2014 (IEEE). http://dx.doi.org/10.1109/NOMS.2014.6838400

21. A. Liotta, D. Constantin Mocanu, V. Menkovski, L. Cagnetta, G. Exarchakos, Instantaneous video quality assessment for lightweight devices, in *Proceedings of the 11th International Conference on Advances in Mobile Computing and Multimedia*, Vienna, Austria, 2–4 Dec 2013 (ACM). http://dx.doi.org/10.1145/2536853.2536903
22. V. Menkovski, A. Liotta, Intelligent control for adaptive video streaming, in *Proceedings of the International Conference on Consumer Electronics*, Las Vegas, US, Jan 11–14, 2013 (IEEE) http://dx.doi.org/10.1109/ICCE.2013.6486825
23. A. Liotta, L. Druda, G. Exarchakos, V. Menkovski, Quality of experience management for video streams: the case of skype, in *Proceedings of the 10th International Conference on Advances in Mobile Computing and Multimedia*, Bali, Indonesia, 3–5 Dec 2012 (ACM) http://dx.doi.org/10.1145/2428955.2428977
24. V. Menkovski, G. Exarchakos, A. Liotta, Machine learning approach for quality of experience aware networks, in *Proceedings of Computational Intelligence in Networks and Systems*, Thessaloniki, Greece, 24–26 Nov 2010 (IEEE)
25. V. Menkovski, G. Exarchakos, A. Liotta, Adaptive testing for video quality assessment, in *Proceedings of Quality of Experience for Multimedia Content Sharing*. Lisbon, Portugal, 29 June 2011 (ACM)
26. G. Exarchakos, V. Menkovski, A. Liotta, Can Skype be used beyond video calling? in *Proceedings of the 9th International Conference on Advances in Mobile Computing and Multimedia*, Ho Chi Minh City, Vietnam, 5–7 Dec 2011 (ACM)
27. V. Menkovski, G. Exarchakos, A. Liotta, Tackling the sheer scale of subjective QoE, in *Proceedings of 7th International ICST Mobile Multimedia Communications Conference*, Cagliari, Italy, 5–7 Sept 2011 (Springer, Lecture Notes of ICST, 2012), vol. 29, pp. 1–15. http://www.springerlink.com/content/p1443m265r25756x/. doi:10.1007/978-3-642-30419-4_1
28. J. Okyere-Benya, M. Aldiabat, V. Menkovski, G. Exarchakos, A. Liotta, Video quality degradation on IPTV networks, in *Proceedings of International Conference on Computing, Networking and Communications*, Maui, Hawaii, USA, Jan 30–Feb 2, 2012 (IEEE)
29. V. Menkovski, G. Exarchakos, A. Liotta, Online QoE prediction, in *Proceedings of the 2nd IEEE International Workshop on Quality of Multimedia Experience*, Trondheim, Norway, June 21–23, 2010 (IEEE)
30. V. Menkovski, G. Exarchakos, A. Liotta, Online learning for quality of experience management, in *Proceedings of The Annual Machine Learning Conference of Belgium and The Netherlands*, Leuven, Belgium, May 27th–28th, 2010. http://dtai.cs.kuleuven.be/events/Benelearn2010/submissions/benelearn2010_submission_20.pdf
31. V. Menkovski, A. Oredope, A. Liotta, A. Cuadra-Sanchez, Optimized online learning for QoE prediction, in *Proceedings of the 21st Benelux Conference on Artificial Intelligence*, Eindhoven, The Netherlands, 29–30 Oct 2009, pp. 169–176. http://wwwis.win.tue.nl/bnaic2009/proc.html, ISSN:1568-7805
32. K. Yaici, A. Liotta, H. Zisimopoulos, T. Sammut, User-centric uality of service management in UMTS, in *Proceedings of the 4th Latin American Network Operations and Management Symposium (LANOMS'05)*, Porto Alegre, Brazil, 29–31 Aug 2005 (Springer)
33. F. Agboma, A. Liotta, User-centric assessment of mobile content delivery, in *Proceedings of the 4th International Conference on Advances in Mobile Computing and Multimedia*, Yogyakarta, Indonesia, 4–6 Dec 2006
34. F. Agboma, A. Liotta, Managing the user's quality of experience, in *Proceedings of the second IEEE/IFIP International Workshop on Business-driven IT Management (BDIM 2007)*. Munich, Germany, 21th May 2007 (IEEE)
35. M. Mu, A. Mauthe, F. Garcia, A utility-based qos model for emerging multimedia applications, in *The Second International Conference on 2008 Next Generation Mobile Applications, Services and Technologies, NGMAST'08* (IEEE, 2008), pp. 521–528
36. M. Mu, R. Gostner, A. Mauthe, G. Tyson, F. Garcia, Visibility of individual packet loss on h.264 encoded video stream–a user study on the impact of packet loss on perceived video quality (2009)

37. J. Kim, T. Um, W. Ryu, B. Lee, Iptv systems, standards and architectures: Part ii-heterogeneous networks and terminal-aware qos/qoe-guaranteed mobile iptv service. Commun. Mag. IEEE **46**(5), 110–117 (2008)
38. M. Volk, J. Sterle, U. Sedlar, A. Kos, An approach to modeling and control of qoe in next generation networks [next generation telco it architectures]. Commun. Mag. IEEE **48**(8), 126–135 (2010)
39. S. Esaki, A. Kurokawa, K. Matsumoto, Overview of the next generation network. NTT Tech. Rev. **5**(6), 2007
40. J. Zhang, N. Ansari, On assuring end-to-end qoe in next generation networks: challenges and a possible solution. Commun. Mag. IEEE **49**(7), 185–191 (2011)
41. R. Fielding, J. Gettys, J. Mogul, H. Frystyk, L. Masinter, P. Leach, T. Berners-Lee, Hypertext transfer protocol–http/1.1 (1999)
42. W. Stevens, G. Wright, *TCP/IP Illustrated: The Protocols*, vol. 1 (Addison-Wesley Professional, 1994)
43. T. Stockhammer, P. Fröjdh, I. Sodagar, S. Rhyu, *Information technologympeg systems technologiespart 6: Dynamic adaptive streaming over http (dash). ISO/IEC, MPEG Draft International Standard* (2011)
44. H. Schulzrinne, Real time streaming protocol (rtsp) (1998)
45. J. Postel, User datagram protocol, *Isi* (1980)
46. P. Frossard, Fec performance in multimedia streaming. Commun. Lett. IEEE **5**(3), 122–124 (2001)
47. V. Menkovski, G. Exarchakos, A. Liotta, A. Cuadra-Sánchez, Managing quality of experience on a commercial mobile tv platform. Int. J. Adv. Telecommun. **4**(1, 2), 72–81 (2011)
48. TM forum—TM forum IPDR program. http://www.tmforum.org/InDepth/10344/home.html
49. A. Cuadra-Sanchez, C. Casas-Caballero, End-to-end quality of service monitoring in convergent iptv platforms, in *Third International Conference on Next Generation Mobile Applications, Services and Technologies, NGMAST'09* (IEEE, 2009), pp. 303–308
50. M. Karam, F. Tobagi, Analysis of the delay and jitter of voice traffic over the internet, in *Proceedings INFOCOM, Twentieth Annual Joint Conference of the IEEE Computer and Communications Societies*, vol. 2 (IEEE, 2001), pp. 824–833
51. E. Alpaydin, *Introduction to machine learning* (MIT press, Cambridge, 2004)
52. J. Quinlan, Induction of decision trees. Mach. Learn. **1**(1), 81–106 (1986)
53. J. Quinlan, *C4. 5: programs for machine learning*, vol. 1 (Morgan Kaufmann, 1993)
54. V. Vapnik, S. Kotz, *Estimation of Dependences Based on Empirical Data*, vol.41 (Springer, New York, 1982)
55. J. Platt et al., Sequential minimal optimization: a fast algorithm for training support vector machines (1998)
56. R. Schapire, The strength of weak learnability. Mach. Learn. **5**(2), 197–227 (1990)
57. L. Breiman, Bagging predictors. Mach. Learn. **24**(2), 123–140 (1996)
58. Y. Freund, R. Schapire et al., Experiments with a new boosting algorithm, in *Machine Learning-International Workshop then Conference* (Morgan Kaufmann Publishers, inc., 1996), pp. 148–156
59. F. Agboma, A. Liotta, QoE-aware QoS management, in *Proceedings of the 6th International Conference on Advances in Mobile Computing and Multimedia*, ser. MoMM '08 (ACM, New York, NY, USA, 2008), p. 111116. http://doi.acm.org/10.1145/1497185.1497210
60. G.T. Fechner, *Elements of Psychophysics* (Holt, Rinehart and Winston, 1966)
61. I. Witten, E. Frank, *Data Mining: Practical Machine Learning Tools and Techniques* (Morgan Kaufmann, 2005)
62. R. Kohavi et al., A study of cross-validation and bootstrap for accuracy estimation and model selection, in *International Joint Conference on Artificial Intelligence*, vol. 14 (Lawrence Erlbaum Associates Ltd, 1995), pp. 1137–1145
63. P. Domingos, G. Hulten, Mining high-speed data streams, in *Proceedings of the Sixth ACM SIGKDD International Conference on Knowledge Discovery and Data Mining* (ACM, 2000), pp. 71–80

64. R. Kohavi, C. Kunz, Option decision trees with majority votes, in *Machine Learning-International Workshop then Conference* (Citeseer, 1997), pp. 161–169
65. B. Pfahringer, G. Holmes, R. Kirkby, New options for hoeffding trees, *AI 2007: Advances in Artificial Intelligence* (2007), pp. 90–99
66. R. Kohavi, Scaling up the accuracy of naive-bayes classifiers: a decision-tree hybrid, in *Proceedings of the Second International Conference on Knowledge Discovery and Data Mining*, vol. 7 (1996)
67. J. Gama, R. Rocha, P. Medas, Accurate decision trees for mining high-speed data streams, in *Proceedings of the Ninth ACM SIGKDD International Conference on Knowledge Discovery and Data Mining* (ACM, 2003), pp. 523–528
68. G. Holmes, R. Kirkby, B. Pfahringer, Stress-testing hoeffding trees. Knowl. Disc. Databases: PKDD **2005**, 495–502 (2005)
69. N. Oza, Online bagging and boosting, in *IEEE International Conference on Systems, Man and Cybernetics*, vol. 3 (IEEE, 2005), pp. 2340–2345
70. N. Oza, S. Russell, Experimental comparisons of online and batch versions of bagging and boosting, in *Proceedings of the Seventh ACM SIGKDD International Conference on Knowledge Discovery and Data Mining* (ACM, 2001), pp. 359–364
71. F. Agboma, A. Liotta, Addressing user expectations in mobile content delivery. Mob. Inf. Syst. **3**(3, 4), 153164 (2007). http://dl.acm.org/citation.cfm?id=1376820.1376823
72. A. Bifet, G. Holmes, B. Pfahringer, R. Kirkby, R. Gavaldà, New ensemble methods for evolving data streams (2009)
73. V. Menkovski, G. Exarchakos, A. Liotta, A. Sánchez, Estimations and remedies for quality of experience in multimedia streaming," in *Third International Conference on Advances in Human-Oriented and Personalized Mechanisms, Technologies and Services (CENTRIC), 2010* (IEEE, 2010), pp. 11–15
74. V. Menkovski, A. Oredope, A. Liotta, A. Sánchez, Predicting quality of experience in multimedia streaming, in *Proceedings of the 7th International Conference on Advances in Mobile Computing and Multimedia* (ACM, 2009), pp. 52–59

Chapter 5
QoE Active Control

Abstract Adaptive networked multimedia services deliver superior overall quality by optimizing a trade-off between the available resources and the delivered quality. Typically this means achieving an efficient system where the maxium amount of users are serviced with the sufficient quality. Successful management also entails adjusting to changes in the environment. Reacting to increase in demand or to the requirements of new devices is essential to meeting customer expectations. However, not all management decisions can be taken reactively. Most services are faced with real-time fluctuations in available resources and real-time requirements that need to be met, which makes real-time adaptation a crucial function. In this chapter we present a framework for QoE aware active control of multimedia services, based on optimal control and reinforcement learning (RL). To demonstrate the applicability of our approach we implement a decision support solution using the suggested approach for the case of adaptive video streaming client.

5.1 QoE Active Control Framework

The QoE active control (QAC) framework implements control solutions for multimedia systems that can be formulated as a Markov Decision Process (MDS) [1]. MDP is a discrete time stochastic control process, which at each iteration is in a state s and decides between a finite number of actions a available in that state. After a specific action is taken, the system transitions to state s' and a corresponding reward or penalty is accumulated. The transition to the next state is not deterministic; and it is associated with a probability distribution. Solving the MDP means determining the optimal action at each state that will maximize the reward or minimize the penalty, all the way until the goal is reached.

The QoE active control framework defines the penalty proportionally to the difference between the maximally achievable QoE and the one actually delivered by the service. Instead of proceeding to model the probabilities in the transition matrix, the framework relies on RL to compute the value of each action a in state s. This way the framework does not need to compute a vast space of conditions and the probabilities for transitioning between them. It rather needs to explore the space of states and

© Springer International Publishing Switzerland 2015
V. Menkovski, *Computational Inference and Control of Quality in Multimedia Services*, Springer Theses, DOI 10.1007/978-3-319-24792-2_5

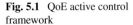
Fig. 5.1 QoE active control framework

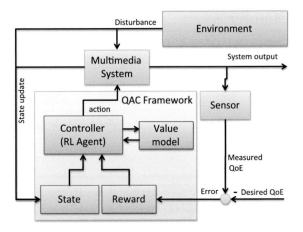

decisions in a process of iterative learning episodes until it discovers the optimal strategy. The state is defined by the condition of the system and the measurement of available resources, and can be partially observable.

Finally, by relying on subjective QoE models to compute the penalty, the framework optimizes the performance of the system in a user centric QoE aware manner. The architecture of the QAC framework is presented in Fig. 5.1.

To demonstrate the applicability of the framework we implement a controller for HTTP adaptive streaming client based on this approach. The following sections describe the HTTP adaptive streaming environment and the existing approaches for control in adaptive streaming. This is followed by a presentation of the intelligent streaming agent, a solution based on our QoE active control framework. Finally, a detailed description of the RL background, the modelling of the penalty function, the state as well as the performance analysis of the agent is presented.

5.2 HTTP Adaptive Streaming Client

For a service to achieve high quality video streaming it needs to accomplish a continuous reproduction of the content with sufficiently high bit-rate and without any errors. However, in best-effort networks multiple sources are competing for the same resources and therefore no guarantees are given that resources will be available when needed. Since video streaming is a data-intensive process, it is particularly susceptible to variations in throughput. If the resources are insufficient the video playback will freeze, unless the video is streamed at a lower bit-rate.

To address the variability in available resources, adaptive streaming technologies are developed such as HTTP streaming [2] and SVC [3]. These technologies allow for continuous adaptation of the bit-rate so that a controlled degradation in quality, or quality of experience (QoE), can be achieved.

Fig. 5.2 HTTP adaptive streaming architecture

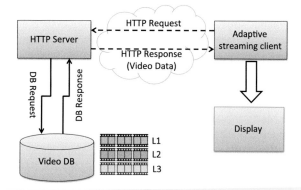

The HTTP adaptive streaming architecture consists of the same high-level compo-nents as other video streaming architectures: servers, transport system and terminal devices. In this particular case the servers can be any typical HTTP web servers. The transport system is implemented over the Internet and the terminal devices are any device that can run the adaptive streaming application (Fig. 5.2).

The video content is transported using the HTTP protocol. The Hypertext transfer protocol (HTTP) is a application-level protocol typically used for transport of web-based content. It is implemented on top of the TCP/IP protocol. HTTP functions in a request-response manner, according to the client-server computing model.

In the case of HTTP adaptive streaming the HTTP servers serve the video content upon receiving a request by a client application. The video content is organized in small segments, or chunks, of few seconds. Different versions of the video with different levels of quality are typically offered. The client can choose to request chunks at specific quality levels (L1, L2, etc.), based on its estimate of the available network throughput and its own control strategy.

5.2.1 DASH Standard

MPEG DASH (Dynamic Adaptive Streaming over HTTP) is a standard for the adap-tive streaming over HTTP [2]. The idea behind standardizing adaptive streaming over HTTP is due to the incompatibilities between proprietary implementation that exist today.

Adaptive streaming is implemented by producing different instances of the source video files with different levels of quality. The main characteristic of this approach is that it uses a HTTP server for delivering the video content. In contrast, other video streaming solutions require dedicated video streaming server applications, which complicate the deployment and management of the service. In addition to the widely understood and deployed HTTP servers, adaptive streaming is further advantaged by the use of HTTP packets, which are firewall friendly and can utilize the HTTP caching mechanisms that already exist in the network.

Apple's HTTP Live streaming, Microsoft's Smooth Streaming and Adobe's HTTP dynamic streaming all use HTTP streaming as their underlying delivery method. Yet each implementation uses its own manifest and segment formats. As the standards vary slightly, the players are not compatible with each other. The goal of the standardization effort is to alleviate this divergence.

The DASH standard defines manifest files that allow the clients to identify the different video streams available. In the standard, the video streams are referred to as Media Presentation and the manifest file is referred to as the Media Presentation Description (MPD).

The MPD is a XML document describing the characteristics of the multimedia content. It has a hierarchical structure [4]. The MPD consist of one or multiple periods, which gave time segments of the multimedia content. Each period consists of one or more adaptation sets. In turn, the adaptation set consist of one or more media components. So one adaptation set contains different bit-rates of the video component, while another adaptation set contains the audio. The media components are defined as representations and consist of multiple segments. The segments are the chunks of media content that the client is requesting from the server.

The client that implements the DASH model needs to first parse the MDP XML file. Next it downloads the representations that are suitable in terms of their characteristics (bit-rate, resolution, and frame-rate) for the available computational and network resources. The client continues to do so as the content is played. The decisions are taken sequentially as the conditions evolve based on the client's control strategy.

The original media description in the MDP file does not contain information about the characteristics of the video itself, such as video SI and TI. These characteristics can help the clients to determine the appropriate subjective QoE models for the content of interest; so we have to incorporate the video characteristics for purposes of QoE management. Subjective QoE models enable more efficient streaming strategies from the point of view of the delivered QoE. Even though the MDP does not specify the use of this type of information, for the implementation of the QAC framework we used extra miscellaneous fields for this purpose.

5.2.2 Existing Heuristic Strategies

As part of the implementation of the QAC framework for adaptive streaming, we started off by evaluating the behaviour of existing clients, through a series of experiments [5]. For the evaluation we implemented a test bed consisting of a HTTP server, an impairment node and a video streaming client (Fig. 5.3).

The test bed consists of a server with Apache HTTP server an impairment node running Linux netm kernel module and the client device. The client device also includes a network monitoring software Wireshark, for observing the client network behaviour. We tested the following streaming platforms: Apple HTTP Live Streaming (HLS); Adobe HTTP Dynamic Streaming (HDS); and Microsoft smooth

Fig. 5.3 Test bed for
adaptive streaming client
evaluation

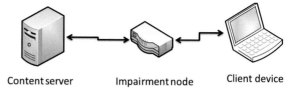

Content server   Impairment node   Client device

streaming (SS). For testing the Microsoft SS, an HTTP module is added to the Apache
HTTP server that offers the Smooth Streaming capabilities. For Adobe HDS and
Microsoft SS a mp4 container is used, while for Apple HLS uses the MPEG-TS container. The video content is 1 min long, compressed on 10 different levels ranging
from 64 kb/s to 2048 kb/s. The videos are segmented in 2 s chunks.

The test is made to determine the QoE performance of adaptation algorithms in
face of various network conditions.

The results of the experiments are illustrated in Fig. 5.4. This figure is a graphical
representation of the sequence with which the video chunks are downloaded. The
horizontal axis represents the video chunk number in the sequence and the vertical
axis represents the bit-rate at which the chunk is downloaded. If a point is present at
$x = 3$ and $y = 4$, then the third chunk is downloaded at $level = 4$ (384 kb/s). The
sequence of downloads is represented with a line connecting the points.

I all cases, all clients start the download at the lowest bit-rate. This is represented
by a point at $(x = 1, y = 1)$. Some players, during playback decide to replace
the already buffered content with content of higher or lower bit-rate. In this case the
figure shows the sequence line going back to a chunk with a lower number at a higher

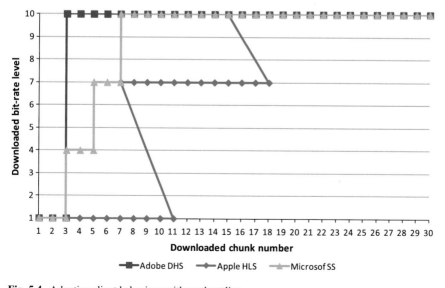

Fig. 5.4 Adaptive client behaviour with no throttling

or lower bit-rate level. This is particularly characteristic for the Apple HLS client. The other two clients, when switching levels only repeat the download of the last video chunk.

When exposed to constant throughput exceeding the maximum bit-rate, Adobe DHS starts off downloading just few chunks at the lowest level and then proceeds to the maximum level. On the other hand Apple HLS downloads more chunks on the lowest bit-rate, than makes one intermediate step and ends up at the highest level. Finally, Microsoft SS gradually increases the quality with two intermediate steps, and finally settling on the maximum level.

When the network throughput is limited to 1 Mb/s (level 8), the Adobe HDS does four cycles from minimum to maximum bit-rate and finally converges on 640 kb/s (Fig. 5.5). The Apple HLS client also does two steps, but converges to only 256 kb/s. Microsoft SS performs much better, in two steps it converges to 768 kb/s.

We simulate high variability in network conditions by shaping the throughput in a square wave (pulse train) fashion. The pulses oscillate between 1 Mb/s (level 8) and 100 kb/s (sufficient for level 1 only) of throughput. The length of the square wave is 10 s, remaining 5 s at each level. The effective throughput in this case, with sufficient buffering (more than 5 s), is 550 kb/s.

In these conditions, Adobe HDS frequently cycles through the bit-rate levels and drops to the lowest one for longer periods (Fig. 5.6). The client cannot find the effective throughput successfully. Furthermore, it reaches buffer depletion and freezes the playback. On the other hands, the Apple HLS client does not cycle, but remains in the low bit-range rate. The Microsoft SS, starts at very low bit-rates

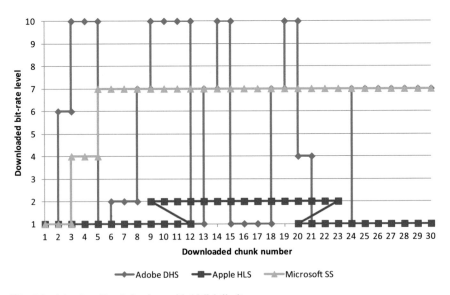

Fig. 5.5 Adaptive client behaviour with 1 Mb/s limit

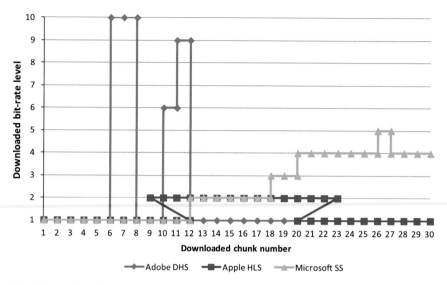

Fig. 5.6 Adaptive client behaviour with changing conditions

and cautiously increases over time. This client finally settled on 384 kb/s, again outperforming the other two clients.

Overall the Adobe HDS makes attempts to improve the QoE by going to higher bit-rates, but ends up causing freezes and frequent quality changes that result in much lower QoE. The Apple HLS avoids freezes, but is very cautious with the bit-rate level. Furthermore, we noticed that it buffers a large amount of the video beforehand. This requires more memory, but is also not friendly to other network users. The Microsoft SS avoids freezes as much as possible, and changes the quality in smoother steps. It strikes a better balance between the bit-rate and the risk of buffer depletion and uses a buffer of no more than 30 s.

In a similar study, the Microsoft SS, the Netflix and the Adobe HDS client were tested [6]. The Microsoft SS is found to be effective under unrestricted and constantly restricted bandwidth. It converges quickly to the maximum available bit-rate and is conservative the quality switching decisions. The Netflix player shows a comparable performance, which is expected since they both use the Microsoft SS platform. The Adobe HDS client does not converge to the appropriate bit-rate even when the throughput is stable. All these findings well aligned to our observations.

In a third study, an evaluation of the changing network conditions in a vehicular environment is implemented [7]. The results show that Microsoft SS achieves the highest average bit-rate and the lowest amount of switching. Apple HLS utilizes the lowest bit-rate and Adobe DHS's performance is the poorest because it introduces freezes.

Overall the three studies reach very similar conclusions. Regardless of the fact that the Microsoft SS performance shows high avoidance of freezes and quality changes, the client does not have any understanding as to how its decisions affect the video QoE. Furthermore, achieving this level of performance without a doubt requires sophisticated heuristics. Adapting them to new devices and content requires significant resources. Instead of this 'design and tuning' approach, we propose an approach of learning or inference, which shortens the development and maintenance time, while providing for high efficiency.

5.3 The Intelligent Streaming Agent

The intelligent streaming agent (ISA) takes a different approach in developing its decision strategy, compared to any of the existing solutions. As discussed in the previous section, the evaluated streaming clients use strategies designed by human experts relying on their intuition and experience. However, such heuristic-based solutions are hard-coded can neither adjust to broad range of conditions, nor adapt as new conditions appear. ISA, on the other hand, infers the optimal strategy by exploration. The inference is guided by the value of each strategy, which is determined by the QoE reward accumulated using that strategy. The QoE reward estimation is based on subjective QoE models.

ISA is an adaptation of the QAC frameworks for HTTP adaptive streaming. It consists of an RL agent for control strategy inference and models for state, action and reward. The models are specifically developed for the particular domain.

The reward function consists of the following factors that affect the QoE in video streaming:

- subjective QoE for the specific video;
- incurred impairments from freezes during playback;
- incurred impairments from change in quality during playback.

The actions are the available bit-rates that the client can choose to download.

The system state captures the conditions of the system that relate to the delivered QoE. These include: the buffer size; buffer utilization; video characteristics; and network performance measurements. The video characteristics include: the size of the video; the current position in the video; resolution; frame-rate; and video SI and TI. The latter two enable the agent to take different actions for different types of video. Since the reward/penalty is calculated using the subjective QoE models, it will differ for different types of video (e.g. content that is low in SI and TI can be compressed more efficiently and requires lower bit-rates). Similarly, the actions need to correspond to the characteristics of the videos.

Based on these three components (state, actions, reward), the agent can now explore and discover the best strategies that optimize the delivered QoE. This inference is implemented using RL training algorithms.

5.3.1 Reinforcement Learning Background

The system under control is issued actions based on the difference between its output and the target or reference output, as defined in control theory. This architecture is well suited when the system output needs to follow a certain reference value over time. However, in some applications a series of actions need to be taken in order to reach a goal or to maintain the system in the desired state. In this case a single action is not of concern, but a strategy to generate sequences or actions is needed.

The purpose of reinforcement learning algorithms is to determine the good strategies that, for a range of specific situations, and as the environment conditions change, will generate actions that achieve the desired system performance.

Supervised learning is not a good approach for this kind of problems, because it is very impractical to generate examples for all the possible conditions that the system can be in and label the best action in each condition. Furthermore, in many cases there is no single best action; rather the sequence of actions determines the performance of the system.

A simple example is given by turn-based games. Many games can be won in many different ways or by different sequences of moves. The success of the play is determined by the end of the game. The value of a single move without the context of the sequence is undetermined.

The reinforcement learning framework addresses this type of problems as an agent operating in a given environment (Fig. 5.7) [8]. The agent takes actions that affect the environment and receives the state and reward according to the changes in the environment. The agent takes decisions on each action from a set of available actions in the current state. After an action has been taken, the state changes accordingly. The reward is calculated based on the changes in the state. The goal of the agent is to optimize its decision in order to accumulate maximum reward.

The environment can be partially observable, so that the state is defined by a probability distribution. Furthermore, the environment can be stochastic, where the actions lead to probabilistic changes in the state.

The inference in RL is implemented by exploring different actions in various states in order to determine the most valuable strategy [9]. Depending on how the reward or penalty function is defined, the best strategy can be: to reach a goal in a minimum number of steps; or to achieve a maximum number of favourable states.

Fig. 5.7 Reinforcement learning framework

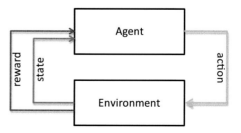

If the agent receives a penalty for each additional step towards the goal, the agent will learn to get to the goal as fast as possible. On the other hand, if reaching certain states on the way is more favourable, the agent will learn take actions accordingly. The agent can even learn to take risks and weigh-in penalties versus rewards for certain actions (e.g. trial-and-error inference).

The learning process is implemented by updating the model for the state-action values. Since the reward is often delayed, the updates for the values need to be propagated back, to actions taken in the past. Different RL algorithms implement the learning in a different manner.

Model based learning is an approach where no exploration is done, but the optimal actions can be determined using dynamic programming [10]. For this approach the state needs to be relatively small and fully observable. This means that the probability $P(s_{t+1}|s_t, a_t)$ of transitioning to state s_{t+1} from state s_t given action a_t needs to be available for all possible transitions and actions. As well as the probability for the received reward $p(r_{t+1}|s_t, s_{t+1})$.

Full understanding of the environment is not usually available, or the cost of computing $P(s_{t+1}|s_t, a_t)$ and $p(r_{t+1}|s_t, s_{t+1})$ is too high for the given state-action space. Hence, the agent needs to explore the environment in order to build the value model for the action-state.

When the agent is exploring, it can observe the value of future states and the reward collected on the way. This information can be used to update the value of the current state. Algorithms that implement this approach are referred to as temporal difference algorithms because they compare the current value of the state (or state-action pair) and compare it to the value of the next state and the reward received.

The online update of the value is implemented by using the delta rule. The simplified version of the update is given in Eq. 5.1, where the system can be in only one state. The value of action a in this state is defined as $Q(a)$. The reward received at time $t + 1$ is r_{t+1}. The value of taking the action at time $t + 1$ (Q_{t+1}) is updated based on the previous value (Q_t) and the received reward (r_{t+1}).

$$Q_{t+1}(a) \leftarrow Q_t(a) + \eta(r_{t+1} - Q_t(a)) \qquad (5.1)$$

η is the learning factor, which is gradually decreased over time for convergence. The convergence occurs when the value of taking the actions is equal to the reward $Q_t(a) \leftarrow r_{t+1}(a)$.

In a more general case, the value of action a in state s is $Q(s, a)$. In this case the system evolves through different states as the actions are taken. Now the value of state-action pair is related to all the rewards that follow after the action has been taken. The update rule is given in Eq. 5.2

$$\hat{Q}(s_t, a_t) \leftarrow \hat{Q}(s_t, a_t) + \eta(r_{t+1} + \gamma \max_{a_t+1} \hat{Q}(s_{t+1}, a_{t+1}) - Q(s_t, a_t)) \qquad (5.2)$$

Instead of just looking at the reward after taking one action, now we need to value the state-action by rewards that are also coming in the following states, hence we take $r_{t+1} + \gamma \max_{a_t+1} \hat{Q}(s_{t+1}, a_{t+1})$, as the value for update. The expression $\max_{a_t} \hat{Q}(s_{t+1}, a_{t+1})$ represents the maximum value that can be achieved with any action. The value of the future states propagated back is discounted by the factor γ, so that the values do not grow to unmanageable sizes as the size of the state space grows.

As the rewards and the state space transitions are probabilistic, the expression $r_{t+1} + \gamma \max_{a_t+1} \hat{Q}(s_{t+1}, a_{t+1})$ is basically a sample from those probabilities. The Q values are now estimates $\hat{Q}$, which converge to the mean values of the probability distributions.

The approach is utilized in the Q-learning algorithm (Algorithm 5.1).

Algorithm 5.1 The Q-Learning algorithm

---
for all episodes **do**
    Initialize s
    repeat
        Choose an action a using $\epsilon - greedy$ exploration
        Take action a, observe r and s'
        $Q(s, a) \leftarrow Q(s, a) + \eta(r + \gamma \arg\max_{a_t+1} Q(s', a') - Q(s, a))$
        $s \leftarrow s'$
    until s is in terminal state
end for

The Q-Learning algorithm always uses the action that produces maximum-valued states to update the current value. This approach is referred to as off-policy learning [9].

However, the action that the agent selects to move to the next state is actually selected by the exploration strategy. The typical exploration strategy is $\epsilon - greedy$. In this strategy the agent selects with probability ϵ uniformly between all the possible actions and with probability $(1 - \epsilon)$ it selects the best known action.

Alternatively on-policy methods use the action chosen by the exploration strategy to update the value. The SARSA method is an example of this approach (Algorithm 5.2).

Algorithm 5.2 Sarsa

---
for all episodes **do**
    Initialize s
    Choose an action a using $\epsilon - greedy$ exploration
    repeat
        Take action a, observe r and s'
        Choose an action a' using $\epsilon - greedy$ exploration
        $Q(s, a) \leftarrow Q(s, a) + \eta(r + \gamma Q(s', a') - Q(s, a))$
        $s \leftarrow s', a \leftarrow a'$
    until s is in terminal state
end for

The current algorithm only updates the previous action/state value, one step in the past. Converging requires passing multiple times over the same states. A way to improve the performance of the RL algorithms is to use Eligibility Traces (ET). ET are records of when the algorithm passed over certain state-actions. Every time the agent at state s takes action a, the trace $e(s, a)$ is set to 1. At the same time all other traces are decayed by $\gamma\lambda$ (Eq. 5.3).

$$e_t(s, a) = \begin{cases} 1 & \text{if } s = s_t \text{ and } a = a_t, \\ \gamma\lambda e_{t-1}(s, a) & \text{otherwise.} \end{cases} \tag{5.3}$$

Now instead of updating just the last visited state we update all the states proportional to their eligibility trace. So, if the state-action is recent in the past the update is more significant, and if the state-action has been visited in more distant past the update is less significant. The temporal difference in SARSA at time t is δ_t (Eq. 5.4).

$$\delta_t = r_{t+1} + \gamma Q(s_{t+1}, a_{t+1}) - Q(s_t, a_t) \tag{5.4}$$

The update with the eligibility traces takes the form given in Eq. 5.5.

$$Q(s, a) \leftarrow Q(s, a) + \eta\delta_t e_t(s, a), \forall s, a \tag{5.5}$$

The λ parameter is the temporal credit. For $\lambda = 0$ the algorithms is updated only one step in the past as in SARSA. The closers λ gets to 1 the longer the updates are made into the past action-states. This algorithm is referred to as SARSA(λ) (Algorithm 5.3).

Algorithm 5.3 SARSA(λ)

Initialize all $Q(s, a)$ arbitrarily, $e(s, a) \leftarrow 0, \forall s, a$
for all episodes **do**
    Initialize s
    Choose an action a using $\epsilon - greedy$ exploration
    repeat
        Take action a, observe r and s'
        Choose an action a' using $\epsilon - greedy$ exploration
        $\delta \leftarrow r + \gamma Q(s', a') - Q(s, a)$
        $e(s, a) \leftarrow 1$
        for all s,a **do**
            $Q(s, a) \leftarrow Q(s, a) + \eta\delta e(s, a)$
            $e(s, a) \leftarrow \gamma\lambda e(s, a)$
            $s \leftarrow s', a \leftarrow a'$
        end for
    until s is in terminal state
end for

However, when the size of the state space starts to grow, the RL algorithms performance starts to deteriorate. In many cases the states can be very similar and the state-actions can have similar values. This means that we can compress the space by making a model that maps the large state space into a smaller one.

One particular example for our case is the state when the available bandwidth supersedes the bit-rate of the highest quality video. In this case we can take the action to download the highest quality, regardless if the bandwidth is twice the bit-rate or ten times the bit-rate. So, these two stats can be mapped into a single state, since the value for the action will be the same. This way the algorithm does not need to visit all possible states to have an approximate value for the actions in that state.

This is a supervised learning approach, where we need to train a model based on the examples available. Now instead of having a table of $Q(s, a)$ for all possible (s, a) pairs, we need a model that maps (s, a) into the value Q ($Q = f(s, a)$). This model can be represented as a linear combination of all the features Φ parameterized by a vector $\vec{\theta}$. The features are binary parameters that characterize the state-action pair. In this case, Q is calculated as given in Eq. 5.6.

$$Q = \sum_{i \in \Phi} \theta(i) \tag{5.6}$$

Instead of updating the values of the table $Q(s, a)$, now we need to updated the parameters $\vec{\theta}$. Since the RL algorithm updates the state online, a suitable approach is to use gradient descent to update the $\vec{\theta}$ parameters. In this method these parameters are update as given in Eq. 5.7.

$$\vec{\theta_{t+1}} = \vec{\theta_t} + \alpha \delta_t \vec{e_t} \tag{5.7}$$

The number of eligibility traces (e_t) now do not correspond to all possible state-action pair, but to all the features that define the (s, a) pair.

Finally to complete the gradient descent SARSA(λ) algorithm we need a mechanism to extract the features from the state-action pairs. Since the features need to be binary this is implemented with a tiling technique [8]. Each discrete parameter in the state-action pair adds a binary digit for each possible value that it can contain. The continual parameters need to be discretized by tiling the space they occupy. The tiles create discrete regions of space. If the value of the parameter falls on a specific tile, its corresponding feature value is 1 or, otherwise, is set to zero. For reasons of efficiency, the continual parameters are joined together in a multidimensional space where the tiles form hyper volumes. This discretization causes loss of fidelity. In order to decrease this effect, more sets of tiles are laid over the parameter space and randomly shifted by different margins, so that they cover slightly different areas.

Our ISA framework implements the gradient-descent SARSA(λ) algorithm for training.

Fig. 5.8 The client
application architecture

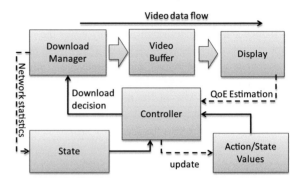

5.3.2 Intelligent Streaming Agent Architecture

The components of our intelligent video client are shown in Fig. 5.8. The download manager downloads the video chunks as indicated by the controller. It also measures network throughput statistics and updates the state model accordingly.

The downloaded video data is stored in the video buffer. The display interface pulls the highest quality video data from the buffer and updates the device display. The display also reports the quality level of the displayed segment to the controller, so that the QoE reward/penalty estimation can be calculated. The controller incorporates a RL agent that continuously selects the optimal decision based on the state of the system and the value for each action at that state. The controller updates the state-action value model based on the reward/penalty accumulated.

The state model contains information about:

- the video spatial and temporal information;
- the video resolution, frame-rate and other objective features;
- the video length;
- the current position in the video stream;
- short term and long term prognosis of the network throughput.

The state is discretized to form a set of binary features using the tiling technique, as described in the previous chapter.

To goal of the agent is to maximize the reward or minimize the penalty. In this architecture the reward is negative and only reaches zero when the highest possible level of quality was achieved. It is calculation is modelled on subjective QoE data.

The controller training implements the linear gradient-descent SARSA(λ) algorithm. Its training is implemented by simulating video streaming playback in a network environment with self-similar background traffic.

Details about the implementation of the reward/penalty calculation, the network throughput prognosis and training performance of the agent are presented in the rest of the chapter.

5.3.3 Reward/Penalty Function

The reward/penalty function for the RL agent calculates the delivered QoE based on the quality of the playback. More precisely the QoE function calculate degradation or a negative reward. The function returns 0 when the playback has encountered no freezes and the video was reproduced with maximum possible bit-rate. As soon as the player chooses to reproduce a lower bit-rate segment, the function returns a negative value. The value is proportional with the degradation given by the subjective MLDS model for the particular type of video. Furthermore, the function includes a quality model for video freezes and for changes in the level of quality. Every time there is a change in quality or a freeze in playback, there is a drop in delivered QoE. Since the changes in quality and the freezes in playback are bursty and localized impairments, their effects on the quality is not constant but a function of time.

The effect of the impairment on the QoE starts at the moment it is introduced. Then it increases in time with a negative gradient. When the impairment stops, the effects on the QoE decay is again with a negative gradient. The next impairment may come before the effects from the previous one have diminished. In this case the effect is cumulative. This type of impairments has a negative effect on QoE, which is proportional to the frequency of their occurrence and their amplitude. Capturing all these effects, we calculate the QoE as given in Eq. 5.8.

$$
\begin{aligned}
QoE &= w_s f_{subjective}(\text{bit-rate}, video_s i, video_t i) \\
&+ w_f f_{freeze}([(t_{p_1}, t_{s_1}), (t_{p_2}, t_{s_2})...(t_{p_n}, t_{s_n})]) + \\
&+ w_l f_{lvlChange}([(\delta_1, t_{l_1}), (\delta_2, t_{l_2})...(\delta_n, t_{l_n})])
\end{aligned}
\tag{5.8}
$$

The $f_{subjective}$ function calculates the degradation due to restrictions in bit-rate. It takes into account the characteristics of the video (spatial and temporal information) and returns a relative value of degradation. A typical subjective quality curve as obtained with the Maximum Likelihood Difference Scaling method [11].

The f_{freeze} function calculates the degradation incurred by the freezes in the playback. The value is based on research done on the psychological effects of this type of impairment. The f_{freeze} inputs a list of pair values. The first (t_{p_i}) is the time at which the ith freeze started and the second (t_{s_i}) is the time at which the playback continued. The effect of the freezes is cumulatively collected over from the beginning to the end of playback as given in Eq. 5.9.

$$
f_{freeze} = \int_0^{t_{end}} I_f(t)dt
\tag{5.9}
$$

The amplitude of the degradation $(I_f(t))$ is proportional to the length of the impairment in time. However, this proportion is not linear. The freeze of 1 s is does not cause half the impairment of a freeze of 2 s. Nor a freeze of 20 s is half as damaging as an impairment of 40 s. The gradient of degradation is high in the beginning and drops over time. If we define the impairment on a relative scale from 0 to 1, we can use an exponential decay function to model the degradation as given in Eq. 5.10 where λ_d is the half life (or the time the quality needs to decrease to half of its initial value). After the freeze has ended and the playback is restarted the impairment remains in the memory of the viewer for a period of time. This period of forgetting or forgiving is also be modelled with a decay function. However, in this case the impairment amplitude is decayed to 0, or the quality rises up to 1. So in the period after the restart of the playback the quality due to freezes can be calculated as given in Eq. 5.10.

$$I_f(t) = \begin{cases} e^{-\frac{t-x_p}{\lambda_d}} & \text{if playback stopped} \\ 1 - e^{-\frac{t-x_s}{\lambda_r}} & \text{if playback continues} \end{cases} \qquad (5.10)$$

where x_p and x_s are the shifts on the time axis and are calculated as given in Eqs. 5.11 and 5.12 respectively.

$$x_p = t_p + \lambda_d \ln(I_f(t_p)) \qquad (5.11)$$

where t_p is the moment the impairment started, and $I_f(t_p)$ is the amplitude of quality at t_p.

$$x_s = t_s + \lambda_r \ln(1 - I_f(t_s)) \qquad (5.12)$$

where t_s is the moment the impairment stopped, and $I_f(t_s)$ is the amplitude of the quality at t_s.

The parameter λ_r is the 'half life' of the repair time after the impairment. Both λ_d and λ_r need to be estimated through subjective studies.

A depiction of the impairment of a freeze during playback is given in Fig. 5.9.

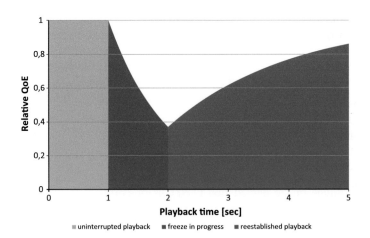

Fig. 5.9 Relative impairment from a freeze

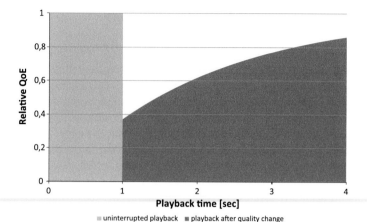

Fig. 5.10 Relative impairment from quality change

The similar approach is taken for the $f_{lvlChange}$ (Eq. 5.13). However, the amplitude of the impairment here is the difference in the distance between levels of quality. And after each occurrence of the change the impairment effect decays to zero (Fig. 5.10).

$$f_{lvlChange} = \int_0^{tend} I_l(t)dt \qquad (5.13)$$

The amplitude of the quality degraded by level change is given in Eq. 5.14.

$$I_l = 1 - e^{-\frac{x - x_l}{\lambda_l}} \qquad (5.14)$$

where λ_l is the half life of the level change impairment decay and x_l is calculated as given in (5.15).

$$x_l = t_l + \lambda_l \ln(1 - I_l(t_l)\frac{N_l - |\delta_l|}{N_l}) \qquad (5.15)$$

where t_l is the moment the impairment happened, and $I_l(t_l)$ is the amplitude of the quality at t_l, N_l is the number of quality levels and δ_l is the difference in quality level between the video that was played before t_l and the video that was played after t_l.

Since the three types of impairments have different impact on QoE, they are weighted differently: w_s, w_f and w_l. These weights can be adjusted by executing a subjective trial and estimating their effect on the QoE.

5.3.4 Estimating Network Throughput Trends

The streaming agent can observe the speed with which the chunks of video are arriving and make assumptions on the available network throughput. This is a passive assessment of available network resources, which requires no additional components and resources. This is why it is a commonly used solution in network streaming clients. Since the network is a shared resource, the available throughput can be highly volatile. However, the resources do not stochastically appear and disappear. The resources are consumed by other services in the network. As these background services (from the point of view of the agent) start and stop, they tend to form certain trends in the available resources.

By looking at the arrival rate of the video data, the agent needs to make assumptions or forecasts about the trends in the available throughput. The accuracy of these forecasts affects the performance of the agent in terms of the delivered QoE.

If the agent had perfect information about the available throughput and could make perfect predictions, we would be dealing with a fully observable system whose actions would be determined via dynamic programming. However, we are dealing with a more challenging situation where future traffic is unknown. We need to "learn" good strategies based on the best estimations available. Estimating trends based on sequences of samples is also referred to as filtering [12]. The challenge in filtering is to overcome the random fluctuations in the data, or the noise, and to detect the trends. However, there is a trade-off between the filtering of high frequency fluctuations and the speed of detecting trends. If the filter is disregarding a wider range of fluctuations it becomes slow to react to trends [13]. On the other hand, if the filter is following the changes in the input too closely, its predictions are short sighted and include the noise. This way average error between the predictions of the filter and the actual measured values is high.

Many strategies to deal with this challenge have been proposed. One of the most popular approaches is the exponentially weighted moving average (EWMA) filter [14]. The filter observes values O_t and outputs the estimations E_t, where E_t is calculated as given in Eq. 5.16.

$$E_t = \alpha E_{t-1} + (1 - \alpha) O_t \qquad (5.16)$$

The α parameter is the smoothing parameter of the filter. High value produce more smoothing, giving a filter that slowly reacts to changes in the data trends. Correspondingly, lower α value make the filter less stable and more agile.

Since both stability and agility are desirable features of the filter in [15] suggest an adaptive approach where the values of α is not constant, but it changes adaptively (5.17).

$$E_t = \alpha_t E_{t-1} + (1 - \alpha_t) O_t \qquad (5.17)$$

The Vertical Horizontal Filter (VHF) solution [13] proposes that the smoothing parameter is computed as in Eq. 5.18.

$$\alpha_t = \beta \frac{\Delta_{max}}{\sum_{i=t-M}^{t} |O_i - O_{i-1}|} \tag{5.18}$$

Δ_{max} is the gap between the maximum and the minimum value in the M most recent observations. β is empirically set to 0.33.

A stability filter dampens the estimates in proportion to the variance of spot observations [16]. The goal is to increase the smoothing when the network exhibits unstable behaviour while keeping the filter stable. On the other hand, keeping low smoothing when the network is more stable results in the filter closely following the trends. To compute the level of instability, this filter uses another EWMA filter as given in Eq. 5.19.

$$U_t = \beta U_{t-1} + (1 - \beta) |O_t - O_{t-1}| \tag{5.19}$$

where $\beta = 0.6$ (selected empirically) and x_t is the value measured at time t. The smoothing parameter α is set as in Eq. 5.20.

$$\alpha_t = \frac{U_t}{U_{max}} \tag{5.20}$$

where U_{max} is the largest instability seen in the 10 most recent observations.

The error based filter is another variation of the adaptive filtering approach, where the gain is adapted according to how well the filter predicts the measurements [16]. When the filter does not accurately predict the future values, the gain is decreased so that the filter estimation will converge more quickly. The error observations are $|E_t - O_t|$. These error observations are filtered through a secondary filter and the estimation error is finally as defined in Eq. 5.21.

$$\Delta_t = \gamma \Delta_{t-1} + (1 - \gamma) |E_{t-1} - O_t| \tag{5.21}$$

where $\gamma = 0.6$ (selected empirically). For the Error based filter the gain is calculated as in Eq. 5.22

$$\alpha_t = 1 - \frac{\Delta_t}{\Delta_{max}} \tag{5.22}$$

where Δ_{max} is the largest instability seen in the 10 most recent observations.

After observing the performance of the filters on the simulated background traffic (Fig. 5.11) we selected a combination of low smoothing (fast) EWMA, high smoothing (slow) EWMA, stability and VHF estimations output as part of the state. In this manner the agent can use the strengths of the different filters to deduce the best strategies when exposed to different traffic patterns.

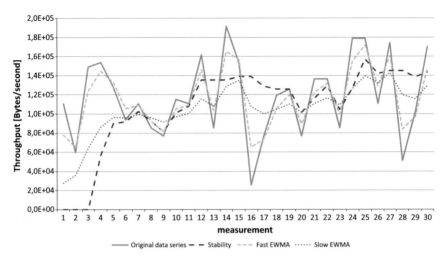

Fig. 5.11 Performance of typical filters

5.3.5 Simulating the Background Traffic

modelling Internet traffic is a lively research area [17]. Different theories propose that
the Internet traffic is self-similar [18]. Measurements also demonstrate self-similarity
of Internet traffic [19]. Self-similarity implies that the traffic distribution is of the
same kind at all time scales. Natural examples of self-similar forms are fractals.
They are geometrically similar over all spatial scales. The Internet traffic, however,
is statistically self-similar over different time scales [19].

Part of the reason for self-similarity lies in the long tailed distribution of file sizes.
Most files are small, very few files are very big. The distribution of file size is long
tailed. Empirically, from file sizes on the world-wide-web (WWW) the distribution
follows a Pareto model [18].

One particular model fits most of these characteristics and is a good fit for modelling Internet traffic. The Poisson Pareto burst process (PPBP) is a simple but accurate traffic model [20].

The length of the bursts of background traffic is distributed with a long-tailed
Pareto distribution. The number of new sources in each iteration is distributed with
a Poisson distribution. An aggregation of the traffic generated of these sources is
self-similar.

The PPBP also has the highly attractive property that its variance-time curve
(the variance of the total traffic arriving in an interval of length t, as a function of
t) is asymptotically, for large t, the same as fractional Brownian noise with Hurst
parameter $H > 0.5$, which is the form that has been observed in real traffic in many
studies [21].

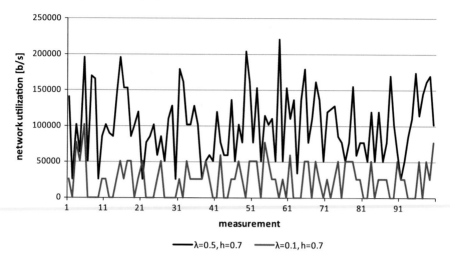

Fig. 5.12 Simulated background traffic with $\lambda = 0.1$ and $\lambda = 0.5$ with $hurst = 0.7$

The number of new processes started in each iteration is drawn from a Poisson distribution (Eq. 5.23).

$$P(x = k) = \frac{\lambda^k e^{-\lambda}}{k!} \tag{5.23}$$

The length of the file downloaded by each process is sampled from a Pareto distribution (5.24).

$$P(X > x) = \begin{cases} \left(\frac{x_m}{x}\right)^\alpha & \text{for } x \geq x_m, \\ 1 & \text{for } x < x_m. \end{cases} \tag{5.24}$$

Example background traffic generated by the PPBP is given in Fig. 5.12.

5.3.6 Agent Performance

We trained the agent in an environment of simulated background traffic and followed the rate of its learning by measuring the delivered QoE after each training episode. The values for the weights of the QoE function were selected as 1 for w_s, 2 for w_l and 10 for w_f. With this selection the penalty for the level change is twice as big as the one for the constant level of quality, and the penalty for a freeze is ten times as big. These values are selected intuitively, based on some simplistic tests. The accurate proportions between the effects of each of the three factors on the QoE need to be established through a comprehensive subjective study.

The agent undertakes a training regimen of 1000 episodes. Over each episode a video streaming session is simulated. Background traffic is generated using the PPBP model, where the *hurst* parameter is set to 0.7 and λ the varies between 0.1

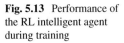

Fig. 5.13 Performance of the RL intelligent agent during training

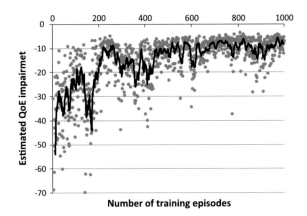

and 1. This creates conditions between very low and very high background traffic. During the training episodes, the penalty is calculated based on the quality level of the played video, the quality changes and the freezes. The penalty incurred over each consecutive episode is given in Fig. 5.13.

The agent learns to avoid freezes quickly, since this is heavily penalized in the QoE function. The results show that the RL agent is fully capable to inferring the appropriate strategy defined by the penalty function, and can be successfully implemented into a video streaming client.

Future developments can possible implement training on network traces in addition to the simulations. This opens possibilities for further exploration of the predictive capabilities on natural patterns in the network.

5.4 Conclusions

Many multimedia services need to include online control mechanisms to optimize their service. Often these control mechanisms are designed with heuristics based on the experience and intuition of the system architects. However, with the growth in complexity of the multimedia systems designing efficient heuristics becomes more difficult. The solution presented in this chapter provides a framework for implementing online control mechanisms that does not require design, but rather infers the optimal strategies [22]. We presented a proof-of-concept implementation of the framework in the form of a HTTP adaptive streaming client to demonstrate the capabilities of this approach.

The implemented solution provides few advances over existing approaches. It relies on subjective models for optimizing the delivered quality. The accuracy of the subjective models in estimating the delivered QoE offer possibility for superior decisions, compared to the existing approaches. There is no design of heuristics involved. The agent is flexible and highly adaptable. It can train and improve continuously.

It infers complex patterns in the traffic and develops appropriate strategies. Finally, since the agent implements autonomic learning, updating the strategy to new content or devices is easily implemented by updating the penalty calculation. The agent in turn will infer the new specific strategies for the novel components. Adding new or more advanced sensors for the network conditions can be implemented by adding more features to the state. There is no need for redesign of new heuristic rules either.

The QAC framework provides a flexible solution to the problem of adaptive streaming. However, it can also be used more generally in other control problems where it is hard to model the system deterministically. Particularly in solutions where human perception (subjective) factors in are key to the performance of the system.

Many researchers are now looking at a number relevant issues to enable active control in adaptive streaming systems. Specific topical issues include: how to automatically measure and possibly control the stream quality in real-time [23–34]; adopt data mining techniques to automate quality assessment [35]; establish appropriate trade-offs between bitrate and received quality [36–40]; find out the sensitivity of quality degradation to video types [41–44]; distribute the stream to increase transmission efficiency [45–48] ; place the streaming severs in optimal locations [49]; enable greater network intelligence [50, 51]; reduce the cost and complexity of subjective studies [52–54]; and develop QoE-aware systems [55–60].

References

1. M. Puterman, *Markov decision processes: Discrete stochastic dynamic programming*. (Wiley, 1994)
2. ISO/IEC 23009-1:2012 - information technology–dynamic adaptive streaming over HTTP (DASH)–part 1: Media presentation description and segment formats. [Online], http://www.iso.org/iso/iso_catalogue/catalogue_tc/catalogue_detail.htm?csnumber=57623
3. G. Van der Auwera, P. David, M. Reisslein, L. Karam, Traffic and quality characterization of the h. 264/avc scalable video coding extension. Adv. Multimed. **2008**(2), 1 (2008)
4. I. Sodagar, The mpeg-dash standard for multimedia streaming over the internet. IEEE Multimedia **18**(4), 62–67 (2011)
5. F. Bertone, V. Menkovski, and A. Liotta, in *Streaming Media with Peer-to-Peer Networks*. Adaptive P2P streaming, IGI Global, 2012, pp. 52–73. [Online], http://www.igi-global.com/chapter/adaptive-p2p-streaming/66305
6. S. Akhshabi, A. Begen, C. Dovrolis, An experimental evaluation of rate-adaptation algorithms in adaptive streaming over http. ACM MMSys **11**, 157–168 (2011)
7. C. Müller, S. Lederer, and C. Timmerer, An evaluation of dynamic adaptive streaming over http in vehicular environments, in *Proceedings of the 4th Workshop on Mobile Video*. ACM, 2012, pp. 37–42
8. R. Sutton and A. Barto, *Reinforcement Learning: An Introduction*, vol. 1, no. 1. (Cambridge Univ Press, 1998)
9. E. Alpaydin, *Introduction to Machine Learning*. (MIT press, 2004)
10. T. Cormen, C. Leiserson, R. Rivest, and C. Stein, *Introduction to Algorithms*. (MIT press, 2001)
11. V. Menkovski and A. Liotta, in *Signal Processing: Image Communication*. Adaptive Psychometric Scaling for Video Quality Assessment, 2012
12. J. Hamilton, *Time series analysis*, vol. 2. (Cambridge Univ Press, 1994)

13. E. Goldoni, G. Rossi, P. Gamba, *Improving Available Bandwidth Estimation Using Averaging Filtering Techniques* (Università degli Studi di Pavia, Laboratorio Reti, Rapport technique, 2008)
14. G. Barnard, Control charts and stochastic processes. J. Roy. Stat. Soc. Ser. B (Methodological), 239–271 (1959)
15. L. Burgstahler and M. Neubauer, New modifications of the exponential moving average algorithm for bandwidth estimation, in *Proceedings of the 15th ITC Specialist Seminar*, 2002
16. M. Kim and B. Noble, Mobile network estimation, in *Proceedings of the 7th annual international conference on Mobile computing and networking*. ACM, 2001, pp. 298–309
17. Z. Sun, D. He, L. Liang, and H. Cruickshank, Internet qos and traffic modelling IET. Software IEE Proc. **151**(5), 248–255 (2004)
18. M. Crovella, A. Bestavros, Self-similarity in world wide web traffic: evidence and possible causes. IEEE/ACM Trans. Netw. **5**(6), 835–846 (1997)
19. W. Leland, M. Taqqu, W. Willinger, D. Wilson, On the self-similar nature of ethernet traffic. ACM SIGCOMM Comput. Commun. Rev. (ACM) **23**(4), 183–193 (1993)
20. M. Zukerman, T. Neame, R. Addie, Internet traffic modeling and future technology implications, in *INFOCOM, Twenty-Second Annual Joint Conference of the IEEE Computer and Communications*. IEEE Societies, vol. 1. IEEE 2003, 587–596 (2003)
21. V. Paxson, S. Floyd, Wide area traffic: the failure of poisson modeling. IEEE/ACM Trans. Netw. (ToN) **3**(3), 226–244 (1995)
22. V. Menkovski, A. Liotta, Intelligent control for adaptive video streaming, in *Proceedings of the International Conference on Consumer Electronics*, Las Vegas, US, January 11–14, 2013 (IEEE), doi:http://dx.doi.org/10.1109/ICCE.2013.6486825
23. M. Torres Vega, E. Giordano, D. C. Mocanu, D. Tjondronegoro, A. Liotta, Cognitive No-reference video quality assessment for mobile streaming services, in *Proceedings of the 7th International Workshop on Quality of Multimedia Experience*, Messinia, Greece, 26–29 May 2015 (IEEE), http://www.qomex.org
24. M. Torres Vega, S. Zou, D. Constantin Mocanu, E. Tangdiongga, A.M.J. Koonen, A. Liotta, End-to-End Performance Evaluation in High-Speed Wireless Networks, in *Proceedings of the 10th International Conference on Network and Service Management*, Rio de Janeiro, Brazil, 17–21 November 2014 (IEEE)
25. D. Constantin Mocanu, G. Santandrea, W. Cerroni, F. Callegati, A. Liotta, Network Performance Assessment with Quality of Experience Benchmarks, in *Proceedings of the 10th International Conference on Network and Service Management*, Rio de Janeiro, Brazil, 17–21 November 2014 (IEEE)
26. D. Constantin Mocanu, G. Exarchakos, H.B. Ammar, A. Liotta, Reduced Reference Image Quality Assessment via Boltzmann Machines, in *Proceedings of the 3rd IEEE/IFIP IM 2015 International Workshop on Quality of Experience Centric Management*, Ottawa, Canada, 11–15 May 2015 (IEEE)
27. M. Torres Vega, D. Constantin Mocanu, R. Barresi, G. Fortino, A. Liotta, Cognitive Streaming on Android Devices, in *Proceedings of the 1st IEEE/IFIP IM 2015 International Workshop on Cognitive Network & Service Management*, Ottawa, Canada, 11–15 May, 2015 (IEEE), http://www.cogman.org
28. V. Menkovski, G. Exarchakos, A. Liotta, A. Cuadra Sánchez, Quality of experience models for multimedia streaming. Int. J. Mobile Comput. Multimed. Commun. **2**(4), 1–20. www.igi-global.com/ijmcmc/, doi:10.4018/jmcmc.2010100101 (ISSN: 1937-9412)
29. A. Liotta, D. Constantin Mocanu, V. Menkovski, L. Cagnetta, G. Exarchakos, Instantaneous video quality assessment for lightweight devices, in *Proceedings of the 11th International Conference on Advances in Mobile Computing and Multimedia (ACM)*, Vienna, Austria, 2–4 Dec 2013, http://dx.doi.org/10.1145/2536853.2536903
30. V. Menkovski, G. Exarchakos, A. Liotta, Online learning for quality of experience management, in *Proceedings of the annual machine learning conference of Belgium and the Netherlands*, Leuven, Belgium, May 27th–28th, 2010. http://dtai.cs.kuleuven.be/events/Benelearn2010/submissions/benelearn2010_submission_20.pdf

31. V. Menkovski, G. Exarchakos, A. Liotta, Online QoE Prediction, in *Proceedings of the 2nd IEEE International Workshop on Quality of Multimedia Experience*, Trondheim, Norway, June 21–23, 2010 (IEEE)

32. V. Menkovski, G. Exarchakos, A. Cuadra-Sanchez, A. Liotta, Measuring Quality of Experience on a commercial mobile TV platform, in *Proceedings of the 2nd International Conference on Advances in Multimedia*, Athens, Greece, 13–19 June 2010 (IEEE)

33. V. Menkovski, A. Oredope, A. Liotta, A. Cuadra-Sanchez, Optimized online learning for QoE prediction, in *Proceedings of the 21st Benelux Conference on Artificial Intelligence*, Eindhoven, The Netherlands, 29–30 Oct 2009 (Open Access: http://wwwis.win.tue.nl/bnaic2009/proc. html, pp. 169–176) (ISSN: 1568-7805)

34. V. Menkovski, A. Oredope, A. Liotta, A. Cuadra-Sanchez, Predicting quality of experience in multimedia streaming, in *Proceedings of the 7th Dec 2009 (ACM)*. http://dl.acm.org/citation. cfm?id=1821766 (ISBN: 978-1-60558-659-5)

35. D.C. Mocanu, G. Exarchakos, A. Liotta, Deep learning for objective quality assessment of 3D images, in *Proceedings of IEEE International Conference on Image Processing*, Paris, France, 27–30 Oct 2014 (IEEE)

36. G. Exarchakos, V. Menkovski, L. Druda, A. Liotta, Network analysis on Skype end-to-end video quality. Int. J. Pervasive Comput. Commun. **11**(1), 2015. http://www.emeraldinsight. com/doi/abs/10.1108/IJPCC-08-2014-0044

37. F. Agboma, A. Liotta, Quality of experience management in mobile content delivery systems. J. Telecommun. Syst. (special issue on the Quality of Experience issues in Multimedia Provision). **49**(1), 85–98 (2012). doi:10.1007/s11235-010-9355-6

38. F. Agboma, A. Liotta, Addressing user expectations in mobile content delivery. J Mobile Inf. Syst. (special issue on Improving Quality of Service in Mobile Information Systems), **3**(3), 153–164 (2007)

39. D.C. Mocanu, A. Liotta, A. Ricci, M. Torres Vega, G. Exarchakos, When does lower bitrate give higher quality in modern video services?, in *Proceedings of the 2nd IEEE/IFIP International Workshop on Quality of Experience Centric Management*, Krakow, Poland, 9 May 2014 (IEEE). doi:http://dx.doi.org/10.1109/NOMS.2014.6838400

40. A. Liotta, L. Druda, G. Exarchakos, V. Menkovski, Quality of experience management for video streams: the case of Skype. in *Proceedings of the 10th International Conference on Advances in Mobile Computing and Multimedia, (ACM)*, Bali, Indonesia, 3–5 Dec 2012. doi: http://dx.doi.org/10.1145/2428955.2428977

41. G. Exarchakos, L. Druda, V. Menkovski, P. Bellavista, A. Liotta, Skype resilience to high motion videos. Int. J. Wavelets Multiresolution Inf. Proc. **11**(3) (2013). doi: http://dx.doi.org/ 10.1142/S021969131350029X

42. V. Menkovski, G. Exarchakos, A. Liotta, The value of relative quality in video delivery, J. Mobile Multimed. **7**(3), 151–162 (2011). ISSN: 1550-4646, http://www.rintonpress.com/ xjmm7/jmm-7-3/151-162.pdf

43. J. Okyere-Benya, M. Aldiabat, V. Menkovski, G. Exarchakos, A. Liotta, Video quality degrada-tion on IPTV networks, in *Proceedings of International Conference on Computing, Networking and Communications*, Maui, Hawaii, USA, Jan 30–Feb 2, 2012 (IEEE)

44. G. Exarchakos, V. Menkovski, A. Liotta, Can Skype be used beyond video calling?, in *Proceed-ings of the 9th International Conference on Advances in Mobile Computing and Multimedia*, Ho Chi Minh City, Vietnam, 5–7 Dec 2011 (ACM)

45. M. Alhaisoni, A. Liotta, M. Ghanbari, Scalable P2P Video Streaming. Int. J. Bus. Data Com-mun. Netw. **6**(3), 49–65 (2010). doi:10.4018/jbdcn.2010070103 (ISSN: 1548-0631)

46. M. Alhaisoni, M. Ghanbari, A. Liotta, Localized multistreams for P2P streaming. Int. J. Digit. Multimed. Broadcast. **843574**, 12 (2010). doi:10.1155/2010/843574

47. M. Alhaisoni, A. Liotta, M. Ghanbari, Resource-awareness and trade-off optimization in P2P video streaming. Int. J. Adv. Media Commun. (special issue on High-Quality Multimedia Streaming in P2P Environments), **4**(1), 59–77 (2010). doi:10.1504/IJAMC.2010.030005 (ISSN 1741-8003)

48. M. Alhaisoni, A. Liotta, Characterization of Signalling and Traffic in Joost. J P2P Netw. Appl. (special issue on Modelling and Applications of Computational P2P), **2**, 75–83 (2009). doi:10. 1007/s12083-008-0015-5 (ISSN: 1936-6450)
49. C. Ragusa, A. Liotta, G. Pavlou, An Adaptive Clustering Approach for the Management of Dynamic Systems. IEEE J. Selected Areas Commun. (JSAC) (special issue on Autonomic Communication Systems), **23**(12), 2223–2235, IEEE (IF 4.138; SJR 3.34) http://dx.doi.org/ 10.1109/JSAC.2005.857203
50. A. Liotta, The Cognitive net is coming. IEEE Spectrum, **50**(8), 26–31, IEEE. doi:http://dx.doi. org/10.1109/MSPEC.2013.6565557
51. A. Liotta, Farewell to deterministic networks, in *Proceedings of the 19th IEEE Symposium on Communications and Vehicular Technology in the Benelux*, Eindhoven, the Netherlands, 16 Nov 2012 (IEEE). doi:http://dx.doi.org/10.1109/SCVT.2012.6399413
52. V. Menkovski, G. Exarchakos, A. Liotta, Tackling the sheer scale of subjective QoE, in *Proceedings of 7th International ICST Mobile Multimedia Communications Conference*, Cagliari, Italy, 5–7 Sep 2011 (Springer, Lecture Notes of ICST, vol. 29, pp. 1–15, 2012). http://www. springerlink.com/content/p1443m265r25756x/, doi:10.1007/978-3-642-30419-4_1
53. V. Menkovski, G. Exarchakos, A. Liotta, Adaptive testing for video quality assessment, in *Proceedings of Quality of Experience for Multimedia Content Sharing*. Lisbon, Portugal, 29 June 2011 (ACM)
54. V. Menkovski, G. Exarchakos, A. Cuadra-Sanchez, A. Liotta, Estimations and remedies for quality of experience in multimedia streaming, in *Proceedings of the 3rd International Conference on Advances in Human-oriented and Personalized Mechanisms, Technologies, and Services*, Nice, France, Aug 22–27, 2010 (IEEE)
55. V. Menkovski, G. Exarchakos, A. Liotta, Machine learning approach for quality of experience aware networks, in *Proceedings of Computational Intelligence in Networks and Systems*, Thessaloniki, Greece, 24–26 Nov 2010 (IEEE)
56. F. Agboma, A. Liotta, QoE-aware QoS management, in *Proceedings of the 6th International Conference on Advances in Mobile Computing and Multimedia*, Linz, Austria, 24–26 Nov 2008
57. F. Agboma, M. Smy, A. Liotta, QoE analysis of a peer-to-peer television system, in *Proceedings of the International Conference on Telecommunications, Networks and Systems*, Amsterdam, Netherlands, 22–24 July 2008
58. F. Agboma, A. Liotta, Managing the user's quality of experience. in *Proceedings of the second IEEE/IFIP International Workshop on Business-driven IT Management (BDIM 2007)*, Munich, Germany, 21 May 2007 (IEEE)
59. F. Agboma, A. Liotta, User-centric assessment of mobile content delivery, in *Proceedings of the 4th International Conference on Advances in Mobile Computing and Multimedia*, Yogyakarta, Indonesia, 4–6 Dec 2006
60. K. Yaici, A. Liotta, H. Zisimopoulos, T. Sammut, User-centric quality of service management in UMTS. in *Proceedings of the 4th Latin American Network Operations and Management Symposium (LANOMS'05)*, Porto Alegre, Brazil, 29–31 Aug 2005 (Springer)

Chapter 6
Conclusions

Abstract Each chapter of this book provides a perspective on key aspects of the given QoE managment proposition. This final chapter, reflects on each of these contributions, their main points and how they tie-together to form an approach for addressing the challenges of QoE management. Here, the main conclusions of each chapter are summarized in the context of the motivation given for the QoE 'learning' management proposition, mainly that the complexity of the multimedia systems is increasing, along with the diversification of content, devices and the evolution in user expectations. Furthermore, it continues to discuss work in the community that builds upon or is closely related to the given proposition. Finally it presents few open questions and directions for the future work in this domain.

6.1 The Principal Message

Multimedia services have become instrumental to how we interact with each other, exchange information, and experience many forms of entertainment. Multimedia content has even entered the domain of printed media, as books and magazines are now augmented with content for tablets and mobile phones. Moreover, the digital versions of these articles includes interactive images and video, blurring the lines between the different modalities.

With ubiquitous connection availability, interactions with these content and services is fully present on mobile devices. We watch movies, TV and user generated content on the move, whether we are at home or in the train. These technologies further enable us to share our experiences and generate content, as mobile devices are not only enabled with connectivity but also with cameras and microphones. We share images, video and even stream what we see to other users, as we see it. This trend shows no signs of slowing down, on the contrary, with developments of wearable computing devices and augmented reality, this trend can only increase. Wearable devices have even more sensors, which allow them to monitor every aspect of our movement, activity and even our physiology . They also allow for recording and displaying of multimedia content. Enriched with all the available information this content can be adapted to the context. Estimating the quality of all these interactions

© Springer International Publishing Switzerland 2015   125
V. Menkovski, *Computational Inference and Control of Quality in Multimedia Services*, Springer Theses, DOI 10.1007/978-3-319-24792-2_6

and new ones that will arise from combining these modalities in yet unforeseen ways would be very hard for a deterministic management system [1].

Even for the types of devices and uses that we know well and are present for a while, such as TV and home cinema, the number of services is growing, as well as the device capabilities and underlying network and display technologies. With a plethora of devices available, adapting the service efficiently is a challenging proposition [2–6]. With standardized technologies, such as SD TV and HDTV, the service parameters are defined precisely. Hence, the delivered quality can be measured more easily since the number of variables is not too high. However, faced with a wide range of devices with different features selecting the appropriate resolution or frame-rate for each possible condition is not trivial. Should the service providers spend all available resources to deliver each pixel of a video to a perfect accuracy in order for the service to have acceptable quality? Is this feasible, and more importantly, is it necessary [7]? A number of studies suggest that, just like human perception, video streaming systems exert a certain degree of resilience, when it comes to quality degradation [8–20]. Yet, getting a more accurate picture is as difficult as it is fascinating.

To address this foundational questions, this book starts with a short discussion on the limitation of the HVS and the masking effects resulting from those limitations [21]. These limitations have been successfully utilized to optimize the quality of multimedia content. Encoding algorithms save on precious bits by disregarding details that are not noticeable by the viewer [22]. Accuracy mechanisms in networks have been substituted for error correction or error concealment mechanisms so that delays are minimized [23]. The specifications for a high quality service have become vaguer. Pushing the limits of technologies to deliver richer features, higher resolution, shorter delays, real-time streaming very often means relaxing some expectations such as reliability or accuracy. Faced with the dilemma of offering more and assuring high quality, service providers are faced with the difficulties of estimating the quality of their service [24–34].

This brings us to QoE as a new metric that addresses the need to measure the quality in these newly developed conditions. However, faced with the complexity of current multimedia systems, the many factors that affect QoE and the continually evolving environment, measuring QoE is a challenge on its own [35].

Chapter 2 addresses the objective measurements of factors that contribute to QoE. QoE is evidently a subjective metric [36], but there are many objective factors that can deliver valuable insight into the level of delivered quality. The results from our evaluation of objective QoE methods demonstrate that, even very simple algorithms, restricted to specific conditions can measure important aspects of quality degradation. On the other hand, complex and sophisticated metrics produce results well correlated with subjective QoE in much wider set of conditions. Understanding how to utilize the objective measurements for evaluating QoE is important for efficient QoE management. Measurement of objective factors is precise, low costs and can be easily automated. Each of these characteristics is important for an efficient QoE management framework [37].

Nevertheless, the key aspect for understanding the QoE is successful subjective quality measurement. Chapter 3 discusses existing subjective QoE methods and their drawbacks [38–40]. In this chapter a new and more effective way for evaluating video QoE via difference scaling, rather than absolute rating, is proposed as a way to achieve accurate subjective video QoE measurement [7]. The MLDS method does not deliver absolute quality ratings, but it provides models that illustrate the utility of the resources against the delivered quality. These models are very well suited for delivering efficient management decisions, as they can be combined with cost functions to derive continuous utility functions [41].

Understanding how to measure objective QoE factors and the delivered subjective QoE is not necessarily enough to guarantee efficient management of a multimedia system. Chapter 4 presents the challenges faced in QoE-based management of a video streaming system. The sheer number of factors involved in this approach practically prohibits the use of efficient subjective QoE modelling [42]. Correlating monitoring data, objective and subjective measurements is necessary to manage the QoE of the service. In this chapter an approach based on Computational Intelligence technologies, as a way to deal with the complexity in determining the highly non-linear relationships among the many monitored parameters and the delivered QoE is proposed. Methods for capturing these relationships into QoE prediction models [43] are discussed. The subjective QoE prediction models provide for estimating the subjective QoE based on objective measurements. Furthermore, online learning solutions to deal with the continuous evolution of technology and expectations in the context of multimedia streaming services [1] is presented. Finally, a method for calculating QoE remedies, as a way to determine which management decisions can deliver satisfactory QoE to the end user [35] follows.

Chapter 5 turns the focus to the real-time management of multimedia systems, where the overall quality is sensitive to the time at which decision are being taken. Typically, video streaming services need to make choices that determine their QoE, based on available resources. Scaling back on bit-rate can mean avoiding a playback freeze that will lower the delivered quality significantly. Commonly in systems such as this, control strategies are designed based on heuristics [44]. However, with the expanding complexity introduced by the various underlying technologies, terminal devices and inevitable changes and upgrades in the environment these solutions often do not provide the best performance and cannot anticipate all operational conditions [45]. In contrast, this work proposes a solution based on 'learning' instead of 'deterministic design' [43, 46]. This solution relies on a reinforcement learning agent to determine the optimal strategies that results in maximum performance. The reinforcement comes from a predefined reward function that guides the inference of the strategy (i.e. parameter values given conditions). Since this reward function is based on fundamental aspects of how the user perceives quality, such as the fidelity, avoiding interruptions and low waiting time, which are not affected by the changes in the technology the algorithm can continuously adapt and learn to improve its performance and function effectively.

To conclude, this work presents a suite of methods and frameworks that address important aspects of QoE in multimedia services. The proposed approach is based on the paradigm of continuous learning and adaptability, rather than deterministic design. In light of the growing complexity and rapid evolution of multimedia services,

computational intelligence methods enable this type of approaches that can address
the given challenges.

Adopting this approach is not without its hurdles. Difference scaling methods offer
important benefits for subjective evaluation, but methods for measuring the subjective
effects of many factors are still missing [41]. Furthermore, the integration of these
models into commercial management systems requires a shift in the management
paradigm from 'How good is the QoE?' to 'How much will resource X improve
the QoE?'. Even though the difference scaling methods lack the 'directness' of the
traditional rating, they more than make up for it with the assessment accuracy [7].

Predictive capabilities, enabled by reinforcement learning, improve the performance of active control agents because they allow for better anticipation of changes
in the environment. However, if reinforcement learning agents are left to continuously learn and adapt to the local environment they will develop unique strategies.
This non-uniformity in deployed products is unusual for service providers. For further adoption of this type of approach an evaluation of long-term performance and
stability needs to be considered.

As with any new approach, a certain level of maturity is necessary for a widespread adoption. Nevertheless, as the current management challenges arising from
the complexity and the diversity continue to grow the continuous learning approach
presents a way out, not only for video enabled services, but also for other multimedia
services.

6.2 Latest Developments

During the development of this work and after its publication [47] many important
results have been published building on it and many other efforts in the area. The
quantity and quality of results speaks of the importance and the impact that this topic
has on multimedia services and management of network services.

New results focus on addressing the estimation of QoE as well as dealing with
the lack of resources and transmission errors for different multimedia services. Estimating QoE for mobile video services has been the topic of [48] and [49], while
Mitra et al. [50] focus on the context in which the service is used as a strong factor in
modelling the QoE. The difficulties of direct subjective estimations and the value of
relative quality estimation proposed in [7] has also been discussed in [51] and [52].
In [53], Kare et al. apply relative assessment on 3D Mobile Video Services as well.
For streaming longer video sequences, which is the standard today, it is important
to look at the different effects on the perceived quality in the longer term. There are
certain effects of adjusting to quality rate, and interruptions that differ from shorter
sequences. In [54], the authors show that both video content and the range of quality
switches significantly influence the end-users' rating behavior. They show that quality level switches are only perceived in high motion sequences or in case switching
occurs between high and low quality video segments. They also confirm that freezes
have significantly higher impact on quality.

In time-sensitive services such as streaming video dealing with loss or corrupted data is one of the key challenges. Because of the latency requiremenets retransmission is typically not possible so it is important to deal with the error in an efficient way. The authors of [55] integrate the characteristics of the video, alignment of the sequence to minimize the effect of the loss on the QoE. On the other hand, pre-emptively lowering the bitrate can in certain situations avoid loss of quality. This strategy could avoid buffer underruns, but also allow for more efficient error correction, avoid signal collisions in wireless transmission and allow more time for retransmission. The analysis of [56] demonstrates such counterintuitive QoS-to-QoE conditions where decreases of quality in a controlled fashion lowers the probability for unexpected problems. This type of effects are further elevated for real-time communications, where the time constraints are even tighter . The work of Exarchakos et al. [57] shows that QoS-based heuristics for control of video quality in applications such as Skype, fail to deal with unexpected changes the environment. High-motion videos are an example of this vulnerability. Such content makes the perceived quality unexpectedly more sensitive to packet inter-arrival time (jitter) than to packet loss. These results suggest that Skype uses if-else heuristics to decide its behavior to QoS changes, which is not always sufficient.

Proactive control is often required in real-time streaming services. Underlying technologies such as HTTP adaptive streaming allow for this kind of approaches. However, these strategies are hard to implement with a heuristic approach. A better approach is based on learning, as proposed in [46] and discussed in Chap. 5. Others in this domain have developed more extensive analysis and proposed methods such as QDASH [58]. The authors of QDASH also present a comparative analysis of different bitrate adaptation strategies in adaptive streaming of monoscopic and stereoscopic video. They implemented subjective testing of the end-user response to the video quality variations. Another method that approaches this problem with reinforcement learning enabled client is given in [59]. This work also presents a comparison with two heuristic methods: the Microsoft IIS Smooth Streaming heuristic and the QoE-driven Rate Adaptation Heuristic for HTTP Adaptive video streaming developed by [60]. They show that the learning-based approach they have implemented has better performance by needing less pre-buffering than the other approaches. The authors of [61] present (Frequency Adjusted) Q-Learning HTTP adaptive streaming client. In contrast to heuristic approaches, their proposed client dynamically learns the optimal behaviour corresponding to the current network environment in order to optimize the QoE. In another study [62], the authors evaluate how the QoE is affected by rapid or slow changes in quality level on different contexts. Finally they also quantify the effect of impairments such as freezes and interruptions to the QoE. These evaluations are important for acquiring a better picture as to how individual aspects of the strategies are affecting the QoE and to design more effective models for evaluating the delivered quality of different control strategies and guiding the inference of these strategies.

In online games, the quality is perceived in a different manner. Here interactivity and responsiveness carry much more importance [63]. For such applications the user is merely interested in the perceived quality, regardless of the underlying network situation. The authors of [63] present an adaptive control mechanism that optimizes the QoE for the use case of a race game, by trading off visual quality against frame rate as a function of the available bandwidth. Their results demonstrate that QoE-driven adaptation leads to improved user experience compared to systems solely focusing on the network QoS.

The overall trend in the domain shows clearly that deterministic approaches are not sufficiently capable to deal with the growth in complexity and variety in networked systems, which underlines the importance of this contribution. In this direction Liotta discusses how as communication networks become increasingly complex and dynamic, key functions known as monitoring, control and management prove to be ineffective [64–66].

6.3 Future Directions

With developments in computer networks new avenues for optimization of QoE of multimedia services are being opened. In the following, a discussion is presented on exciting new technologies and how these create new dimensions for QoE management.

Next-generation networks where voice, data, and multimedia services are planned to be converged onto a single network platform with increasing complexity and heterogeneity of underlying wireless and optical networking systems. However, challenges relating to optimization of QoE are not necessarily dealt with by the underlying technologies. In [67] the authors discuss challenges and a possible solution for optimizing end-to-end QoE in Next-Generation Networks.

Software defined networks enable rapid development of services, reducing IT costs and improving workflow efficiency. However, these systems also create a new dimension of interaction for the QoE of multimedia services operating in this environment. In [68], Mocanu et al. introduce a new network performance methodology based on QoE benchmarks for video streaming services. They show how to evaluate quality degradation in software defined networks. Their approach is suited for the evaluation of dynamic networks and helps to better pinpointing the critical factors that affect the applications the most.

Peer-to-peer (P2P) video streaming has been an topic of research for a while, promising an approach that deals with scalability of these services. Results from [69] show that a well-defined behaviour of a parent selection software agent can improve the continuity in the video streaming experience. They also show that this peer welcoming behaviour of the agents decreases the internal Internet service provider traffic significantly.

As discussed in Chap. 3, modelling QoE is typically expensive because it requires subjective examinations. It would be very efficient if QoE could be estimated based on the behaviour of the user. This is commonly hard to implement. The work presented in [70] however, demonstrates an approach for estimating QoE of a HTTP video streaming from the user-viewing activities. In this work they analyze user-viewing activities and correlate them with network path performance and the QoE. The results show that network measurement alone do not convey important information about the perceived video quality. Moreover, since video impairments can trigger user-viewing activities, notably pausing and reducing the screen size, including these events into the prediction model, can increase the model accuracy.

Finally one important aspect for learning methods for QoE management are developments in the Machine Learning domain. These developements particularly in Deep Learning models allow for modelling high level abstractions that could be used for improving the accuracy of QoE models. These methods are commonly based on developing deep artificial neural network models. One study that works in this direction is developed by Dent et al. in [71]. In this work Deng et al. also focus on transferring the knowledge of the model from one context to another by feature transformation. The work of Mocanu et al. in [72] uses deep learning to make predictions on the user's QoE on 3D images using Factored Third Order Restricted Boltzmann Machine (Q3D-RBM). They develop a Reduced Reference QoE assessment process for automatic image assessment and anticipate potential in applying their work to 3D video assessment.

In the real-time QoE control domain Deep Learning could also have strong influence. Developments in Deep Reinforcement Learning [73] open possibilities for leveraging the highly effective feature learning capabilities of these models to develop a much more efficient state space representation for reinforcement learning agents.

These components show important advances and future directions for the development of QoE management approaches that deal with the challenges in the highly connected global environment.

References

1. V. Menkovski, A. Oredope, A. Liotta, A. Sánchez, Predicting quality of experience in multimedia streaming, in *Proceedings of the 7th International Conference on Advances in Mobile Computing and Multimedia*. ACM, 2009, pp. 52–59
2. F. Agboma, A. Liotta, Addressing user expectations in mobile content delivery. J. Mob. Inf. Syst. (special issue on Improving Quality of Service in Mobile Information Systems) **3**(3), 153–164 (IOS Press, 2007)
3. M. Alhaisoni, A. Liotta, Characterization of signalling and traffic in joost. J. P2P Netw. Appl. (special issue on Modelling and Applications of Computational P2P) **2** 75–83, (Springer, 2009). doi:10.1007/s12083-008-0015-5, ISSN:1936-6450
4. M. Alhaisoni, A. Liotta, M. Ghanbari, Resource-awareness and trade-off optimization in P2P video streaming. Int. J. Adv. Media Commun. (special issue on High-Quality Multimedia Streaming in P2P Environments) **4**(1), 59–77 (Inderscience Publishers, 2010) doi:10.1504/IJAMC.2010.030005 ISSN:1741-8003
5. M. Alhaisoni, M. Ghanbari, A. Liotta, Localized multistreams for P2P streaming. Int. J. Digit. Multimedia Broadcast. **843574**, 12. Hindawi 2010, (2010). doi:10.1155/2010/843574
6. M. Alhaisoni, A. Liotta, M. Ghanbari, Scalable P2P video streaming, Int. J. Bus. Data Commun. Netw. **6**(3), 49–65 (IGI Global, 2010) doi:10.4018/jbdcn.2010070103 ISSN:1548-0631
7. V. Menkovski, G. Exarchakos, A. Liotta, The value of relative quality in video delivery. J. Mob. Multimedia **7**(3), 151–162 (2011)

8. G. Exarchakos, L. Druda, V. Menkovski, P. Bellavista, A. Liotta, Skype resilience to high motion videos. Int. J. Wavelets, Multiresolut. Inf. Process. **11**(3), 2013 (World Scientific Publishing). http://dx.doi.org/10.1142/S021969131350029X

9. F. Agboma, A. Liotta, Quality of experience management in mobile content delivery systems, J. Telecommun. Syst. (special issue on the Quality of Experience issues in Multimedia Provision) **49**(1), 85-98 (Springer, 2012). doi:10.1007/s11235-010-9355-6

10. V. Menkovski, G. Exarchakos, A. Liotta, A. Cuadra Sánchez, Quality of experience models for Multimedia Streaming. Int. J. Mob. Comput. Multimedia Commun. **2**(4), 1–20 (IGI Global, Oct–Dec 2010). doi:10.4018/jmcmc.2010100101 ISSN:1937-9412 (www.igi-global.com/ijmcmc/)

11. M. Torres Vega, S. Zou, D. Constantin Mocanu, E. Tangdiongga, A.M.J. Koonen, A. Liotta, End-to-end performance evaluation in high-speed wireless networks, in *proceedings of the 10th International Conference on Network and Service Management* (Rio de Janeiro, Brazil, 17–21 Nov 2014)

12. A. Liotta, L. Druda, G. Exarchakos, V. Menkovski, Quality of experience management for video streams: the case of skype, in *proceedings of the 10th International Conference on Advances in Mobile Computing and Multimedia*, Bali, Indonesia, 3–5 Dec 2012 (ACM). http://dx.doi.org/10.1145/2428955.2428977

13. G. Exarchakos, V. Menkovski, A. Liotta, Can skype be used beyond video calling?, in *proceedings of the 9th International Conference on Advances in Mobile Computing and Multimedia*, Ho Chi Minh City (Vietnam, 5–7 Dec 2011) (ACM)

14. V. Menkovski, G. Exarchakos, A. Liotta, Tackling the sheer scale of subjective QoE, in *Proceedings of 7th International ICST Mobile Multimedia Communications Conference*, Cagliari, Italy, 5–7 Sept 2011 (Springer, Lecture Notes of ICST, 2012), vol. 29, pp. 1–15 doi:10.1007/978-3-642-30419-4_1. http://www.springerlink.com/content/p1443m265r25756x/

15. J. Okyere-Benya, M. Aldiabat, V. Menkovski, G. Exarchakos, A. Liotta, Video quality degradation on IPTV networks, in *Proceedings of International Conference on Computing, Networking and Communications*, Maui, Hawaii, USA, Jan 30–Feb 2, 2012 (IEEE)

16. K. Yaici, A. Liotta, H. Zisimopoulos, T. Sammut, User-centric quality of service management in UMTS, in *Proceedings of the 4th Latin American Network Operations and Management Symposium (LANOMS'05)*, Porto Alegre, Brazil, 29–31 Aug 2005 (Springer)

17. F. Agboma, A. Liotta, User-centric assessment of mobile content delivery, in *Proceedings of the 4th International Conference on Advances in Mobile Computing and Multimedia*, (Yogyakarta, Indonesia, 4–6 Dec 2006)

18. F. Agboma, A. Liotta, Managing the user's quality of experience. in *Proceedings of the second IEEE/IFIP International Workshop on Business-driven IT Management (BDIM 2007)*. Munich, Germany, 21th May 2007 (IEEE)

19. F. Agboma, M. Smy, A. Liotta, QoE analysis of a peer-to-peer television system, in *Proceedings of the International Conference on Telecommunications, Networks and Systems* (Amsterdam, Netherlands, 22–24 July 2008)

20. F. Agboma, A. Liotta, QoE-aware QoS management, in *Proceedings of the 6th International Conference on Advances in Mobile Computing and Multimedia* (Linz, Austria, 24–26 Nov 2008)

21. W. Hendee, P. Wells, *The perception of visual information* (Springer, 1997)

22. J. Li, Visual progressive coding, in *SPIE proceedings series*. Society of Photo-Optical Instrumentation Engineers, 1998, pp. 1143–1154

23. A. Watson, L. Kreslake, Measurement of visual impairment scales for digital video, in *Proceedings of SPIE Human Vision, Visual Processing, and Digital Display*, vol. 4299 (2001)

24. V. Menkovski, G. Exarchakos, A. Liotta, A. Sánchez, Measuring quality of experience on a commercial mobile tv platform, in *IEEE Second International Conferences on Advances in Multimedia (MMEDIA)*, pp. 33–38 (2010)

25. V. Menkovski, G. Exarchakos, A. Liotta, Online QoE prediction, in *Proceedings of the 2nd IEEE International Workshop on Quality of Multimedia Experience*, Trondheim, Norway, June 21–23, 2010 (IEEE)

26. V. Menkovski, A. Oredope, A. Liotta, A. Cuadra-Sanchez, Optimized online learning for QoE prediction, in *Proceedings of the 21st Benelux Conference on Artificial Intelligence*, Eindhoven, The Netherlands, 29–30 Oct 2009. http://wwwis.win.tue.nl/bnaic2009/proc.html, 169–176, ISSN:1568-7805

27. V. Menkovski, G. Exarchakos, A. Liotta, Online learning for quality of experience management, in *Proceedings of the annual machine learning conference of Belgium and The Netherlands*, Leuven, Belgium, May 27th–28th, 2010. http://dtai.cs.kuleuven.be/events/Benelearn2010/submissions/benelearn2010_submission_20.pdf

28. V. Menkovski, A. Liotta, Adaptive psychometric scaling for video quality assessment. J. Signal Process.: Image Commun. **26**(8), 788–799 (Elsevier, 2012). http://dx.doi.org/10.1016/j.image.2012.01.004

29. V. Menkovski, G. Exarchakos, A. Liotta, Machine learning approach for quality of experience aware networks, in *Proceedings of Computational Intelligence in Networks and Systems*, Thessaloniki, Greece, 24–26 Nov 2010 (IEEE)

30. V. Menkovski, G. Exarchakos, A. Liotta, Adaptive testing for video quality assessment, in *Proceedings of Quality of Experience for Multimedia Content Sharing*, Lisbon, Portugal, 29 June 2011 (ACM)

31. A. Liotta, D. Constantin Mocanu, V. Menkovski, L. Cagnetta, G. Exarchakos, Instantaneous video quality assessment for lightweight devices, in *Proceedings of the 11th International Conference on Advances in Mobile Computing and Multimedia*, Vienna, Austria, 2–4 Dec 2013 (ACM). http://dx.doi.org/10.1145/2536853.2536903

32. D. Constantin Mocanu, G. Exarchakos, H.B. Ammar, A. Liotta, Reduced reference image quality assessment via Boltzmann machines, in *Proceedings of the 3rd IEEE/IFIP IM 2015 International Workshop on Quality of Experience Centric Management*, Ottawa, Canada, 11–15 May 2015 (IEEE)

33. M. Torres Vega, D. Constantin Mocanu, R. Barresi, G. Fortino, A. Liotta, Cognitive streaming on android devices, in *Proceedings of the 1st IEEE/IFIP IM 2015 International Workshop on Cognitive Network & Service Management*, Ottawa, Canada, 11–15 May, 2015 (IEEE). http://www.cogman.org

34. M. Torres Vega, E. Giordano, D. C. Mocanu, D. Tjondronegoro, A. Liotta, Cognitive no-reference video quality assessment for mobile streaming services, in *Proceedings of the 7th International Workshop on Quality of Multimedia Experience*, Messinia, Greece, 26–29 May 2015 (IEEE). http://www.qomex.org

35. V. Menkovski, G. Exarchakos, A. Liotta, A. Sánchez, Estimations and remedies for quality of experience in multimedia streaming, in *IEEE Third International Conference on Advances in Human-Oriented and Personalized Mechanisms, Technologies and Services (CENTRIC)*, 2010, 2010, pp. 11–15

36. S. Winkler, On the properties of subjective ratings in video quality experiments, in *IEEE International Workshop on Quality of Multimedia Experience, QoMEx 2009*, pp. 139–144 (2009)

37. S. Winkler, P. Mohandas, The evolution of video quality measurement: from PSNR to hybrid metrics. IEEE Trans. Broadcast. **54**(3), 660668 (2008). http://ieeexplore.ieee.org/xpls/abs_all.jsp?arnumber=4550731

38. K. Seshadrinathan, R. Soundararajan, A. Bovik, L. Cormack, A subjective study to evaluate video quality assessment algorithms, in *SPIE Proceedings Human Vision and Electronic Imaging*, vol. 7527 (Citeseer, 2010)

39. P. Corriveau, C. Gojmerac, B. Hughes, L. Stelmach, All subjective scales are not created equal: the effects of context on different scales. Signal Process **77**(1), 1–9 (1999)

40. V. Menkovski, G. Exarchakos, A. Liotta, Tackling the sheer scale of subjective QoE, in *Mobile Multimedia Communications* (Springer, 2012) pp. 1–15

41. V. Menkovski, A. Liotta, Adaptive psychometric scaling for video quality assessment, *Signal Processing: Image Communication*, 2012

42. V. Menkovski, A. Liott, QoE for mobile streaming, in *Mobile Multimedia—User and Technology Perspectives*, ed. by D. Tjondronegoro (InTech, 2012). http://www.intechopen.com/books/mobile-multimedia-user-and-technology-perspectives/qoe-for-mobile-streaming

43. V. Menkovski, G. Exarchakos, A. Liotta, A. C. Sánchez, Quality of experience models for multimedia streaming, in *Advancing the Next-Generation of Mobile Computing: Emerging Technologies: Emerging Technologies*, p. 112, 2012
44. J. Chakareski, B. Girod, Rate-distortion optimized video streaming over internet packet traces, in *IEEE International Conference on Image Processing, 2005. ICIP 2005*, vol. 2 (2005), pp. II–161
45. D. Jarnikov, T. Özçelebi, Client intelligence for adaptive streaming solutions. Signal Process.: Image Commun. **26**(7), 378–389 (2011)
46. V. Menkovski, A. Liotta, Intelligent control for adaptive video streaming, in *Proceedings of the international conference on consumer electronics* (Las Vegas, 2013)
47. V. Menkovski, Computational inference and control of quality in multimedia services, Ph.D. dissertation, Ph.D. Thesis, University of Eindhoven, 2013. ISBN: 978-90-386-3355-8
48. W. Song, D. W. Tjondronegoro, M. Docherty, in *Understanding user experience of mobile video: framework, measurement, and optimization* (INTECH Open Access Publisher, 2012)
49. W. Song, D.W. Tjondronegoro, Acceptability-based qoe models for mobile video. IEEE Trans. Multimedia **16**(3), 738–750 (2014)
50. K. Mitra, A. Zaslavsky, C. Ahlund, Context-aware QoE modelling, measurement and prediction in mobile computing systems (2013)
51. S. Ickin, M. Fiedler, K. Wac, P. Arlos, C. Temiz, K. Mkocha, Electronic research archive of blekinge institute of technology (2014)
52. S. Ickin, M. Fiedler, K. Wac, P. Arlos, C. Temiz, K. Mkocha, VLQoE: Video QoE instrumentation on the smartphone, in *Multimedia Tools and Applications*, pp. 1–31, 2014
53. P. A. Kara, L. Bokor, and S. Imre, Analysis of assessment alteration phenomena of subjective quality of experience measurements in 2d and 3d mobile video services (2014)
54. N. Staelens, J. De Meulenaere, M. Claeys, G. Van Wallendael, W. Van den Broeck, J. De Cock, R. Van de Walle, P. Demeester, F. De Turck, *Subjective Quality Assessment of Longer Duration Video Sequences Delivered Over http Adaptive Streaming to Tablet Devices* (2014)
55. F. Surez, A. Garca, J. Granda, D. Garca, and P. Nuo, Assessing the QoE in video services over lossy networks. J. Netw. Syst. Manag. 1–24 (2015). http://dx.doi.org/10.1007/s10922-015-9343-y
56. D. C. Mocanu, A. Liotta, A. Ricci, M. T. Vega, G. Exarchakos, When does lower bitrate give higher quality in modern video services?, in *IEEE Network Operations and Management Symposium (NOMS), 2014* (2014) pp. 1–5
57. G. Exarchakos, L. Druda, V. Menkovski, A. Liotta, I. Khalil, I. Khalil, Network analysis on skype end-to-end video quality. Int. J. Pervasive Comput. Commun. **11**(1) (2015)
58. R. K. Mok, X. Luo, E. W. Chan, R. K. Chang, Qdash: a QoE-aware dash system, in *Proceedings of the 3rd Multimedia Systems Conference* (ACM, 2012) pp. 11–22
59. J. van der Hooft, S. Petrangeli, M. Claeys, J. Famaey, F. De Turck, A learning-based algorithm for improved bandwidth-awareness of adaptive streaming clients (2015), pp. 131–138
60. S. Petrangeli, M. Claeys, S. Latré, J. Famaey, and F. De Turck, A multi-agent q-learning-based framework for achieving fairness in http adaptive streaming, in *IEEE Network Operations and Management Symposium (NOMS), 2014* (2014), pp. 1–9
61. M. Claeys, S. Latré, J. Famaey, T. Wu, W. Van Leekwijck, F. De Turck, Design and optimisation of a (fa) q-learning-based http adaptive streaming client. Connection Sci. **26**(1), 25–43 (2014)
62. S. Tavakoli, J. Gutiérrez, N. Garcia, Subjective quality study of adaptive streaming of monoscopic and stereoscopic video. IEEE J. Sel. Areas Commun. **32**(4), 684–692 (2014)
63. B. Vankeirsbilck, T. Verbelen, D. Verslype, N. Staelens, F. De Turck, P. Demeester, B. Dhoedt, Quality of experience driven control of interactive media stream parameters, in *IEEE International Symposium on Integrated Network Management IFIP/IEEE (IM, 2013)*, pp. 1282–1287 (2013)
64. A. Liotta, Farewell to deterministic networks, in *IEEE 19th Symposium on Communications and Vehicular Technology in the Benelux (SCVT), 2012*, pp. 1–4 (2012)
65. A. Liotta, The cognitive net is coming, IEEE Spectr. **50**(8), 26–31, Aug 2013 (IEEE). http://dx.doi.org/10.1109/MSPEC.2013.6565557

66. C. Ragusa, A. Liotta, G. Pavlou, An adaptive clustering approach for the management of dynamic systems. IEEE J. Sel. Areas Commun. (JSAC), (special issue on Autonomic Communication Systems) **23**(12), 2223–2235, IEEE, Dec 2005 (IF 4.138; SJR 3.34). http://dx.doi.org/10.1109/JSAC.2005.857203

67. J. Zhang, N. Ansari, On assuring end-to-end QoE in next generation networks: challenges and a possible solution. IEEE Commun. Mag. **49**(7), 185–191 (2011)

68. D. C. Mocanu, G. Santandrea, W. Cerroni, F. Callegati, A. Liotta, Network performance assessment with quality of experience benchmarks, in *IEEE 10th International Conference on Network and Service Management (CNSM), 2014* (2014), pp. 332–335

69. K.D. Teket, M. Sayit, G. Kardas, Software agents for peer-to-peer video streaming. IET Softw. **8**(4), 184–192 (2014)

70. R. K. Mok, E. W. Chan, X. Luo, R. K. Chang, Inferring the qoe of http video streaming from user-viewing activities, in *Proceedings of the first ACM SIGCOMM workshop on Measurements up the stack*, (ACM, 2011), pp. 31–36

71. J. Deng, L. Zhang, J. Hu, D. He, *Adaptation of Ann Based Video Stream Qoe Prediction Model*. Advances Multimedia Information Processing-PCM (Springer, 2014), pp. 313–322

72. D. C. Mocanu, G. Exarchakos, A. Liotta, Deep learning for objective quality assessment of 3d images (IEEE Service Center, 2014)

73. V. Mnih, K. Kavukcuoglu, D. Silver, A.A. Rusu, J. Veness, M.G. Bellemare, A. Graves, M. Riedmiller, A.K. Fidjeland, G. Ostrovski et al., Human-level control through deep reinforcement learning. Nature **518**(7540), 529–533 (2015)

Printed in the United States
By Bookmasters